NATIONAL GEOGRAPHIC

POCKET GUIDE TO THE
Night Sky
OF NORTH AMERICA

CATHERINE HERBERT HOWELL

NATIONAL GEOGRAPHIC

WASHINGTON, D.C.

CONTENTS

Introduction 6

||||||||||||||||||||||||||||||||||||||||||||||||||||||||||||||||||||||

||||||||||||||||||||||||||||||||||||||||||||||||||||||||||||||||||||||

III

Night Sky
Nature's Nocturnal Entertainment

We tend to overlook the night sky, because nighttime itself is something we work around; often nothing but a light switch stands between our day and night activities. But the night sky is worth knowing: It is our window on the past, present, and future. Knowledge of night sky phenomena—moon, stars, planets, comets, and galaxies—draws us in toward the origins of the universe, helps us understand the flow of the seasons, and makes us ponder our planet's destiny. Above all (no pun intended), observing the night sky is the ultimate free pastime, available to all without a reservation.

What's Out There

So what am I looking at? Are those little lights the stars, planets, planes, meteors, or what? Where are those gods, goddesses, and animals with their connect-the-dots outlines? There are no outlines, just stars. But with time and practice, you can discern the shapes of many constellations and learn to tell apart stars and satellites, comets and spacecraft, planets and airplanes.

Stars move from east to west through the night. They twinkle because their pinpoints of light are perturbed by fluctuations in the atmosphere. Planets don't usually twinkle because they are much closer to us. Unlike starlight, planets' disk shapes reflect light from the sun, and this relatively broader width of light is less affected by the atmosphere. Five planets are visible to the naked eye: Mercury, Venus, Mars, Jupiter, and Saturn. Mercury is a bit elusive, but it can be viewed well at certain intervals during the year. Venus is the second brightest object in the night sky after the moon. It shines a brilliant white. Mars glows orangish red, while Jupiter is a steady white, and Saturn is pale yellow. Planets take the same path that the sun and moon take across the sky.

The blurry band of light running across the sky, high in winter and summer, is the Milky Way. Yes, we also reside in the Milky Way, but we are able to view other parts of it. Small blurry blobs in the sky are star

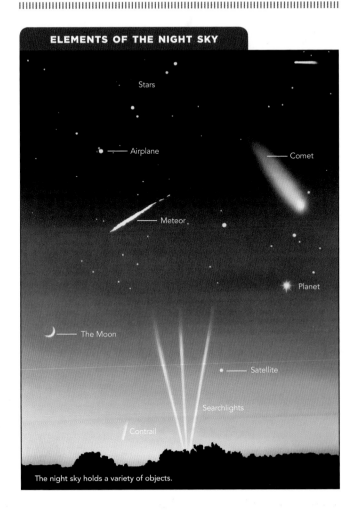

ELEMENTS OF THE NIGHT SKY

Stars

Airplane

Comet

Meteor

Planet

The Moon

Satellite

Searchlights

Contrail

The night sky holds a variety of objects.

clusters, nebulae, or perhaps even a galaxy, such as the Andromeda. It appears as a little blur in the constellation Andromeda and is the most distant object visible to the naked eye.

Fast-moving objects include meteors (shooting stars), which are bits of solar system debris that flame out in Earth's atmosphere. Slower and steadily moving objects with steady, blinking lights are airplanes.

Satellites are visible just after sunset or before sunrise, or during the night depending on latitude; they appear as small, starlike lights that move smoothly before disappearing into Earth's shadow.

Getting Oriented

Earthly reference points are familiar—the marks on a compass; the Equator dividing the planet into hemispheres; the North and South Poles; measurements of longitude and latitude to fix location. Astronomical conventions for locating objects in the sky use similar terms. Applying them requires both a conceptual leap and an appreciation of the fact that what you see and where it appears depend on where you are and when you are looking.

Finding North

In the Northern Hemisphere, the star Polaris is the starting point. Located almost directly above Earth's North Pole, it is visible all year. Find it by spotting the Big Dipper and extending an imaginary line through the two stars at the far edge of its "bowl." Polaris is approximately one "dipper's length" away. Face it, and you face north— south is behind you, west on the left, and east on the right.

Polaris's height in the sky also indicates latitude. At the North Pole (90° N), the star is directly overhead at the highest point in the sky—a 90-degree angle to the horizon. Moving south, the star drops lower in the sky. From the northern United States and southern Canada, corresponding to latitudes of around 40° to 50° N, Polaris is roughly halfway up the northern sky.

The Celestial Sphere

It is now known that Earth is not at the center of the cosmos, but star maps are still oriented as if it were. It will help, in fact, to imagine that Earth is enveloped by a giant globe called the celestial sphere. Its equator—the celestial equator—is parallel to Earth's. Lines extending out from Earth's North and South Poles pass through the sphere's north and south celestial poles.

Next, visualize that all the stars are attached to this sphere. Stars can be found using two coordinates called declination and right

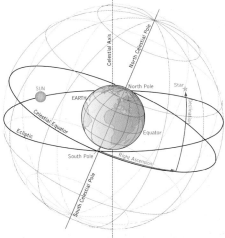

ascension. The first is akin to latitude—how far the star is above or below the celestial equator, measured from zero (at the equator) to plus (for the north) or minus (for the south) 90°. Right ascension is akin to longitude. It is divided, just like Earth's day, into 24 hours, with each hour equaling 15° of circumference. The starting "zero hour" is a north-south line that marks where the sun hits the Earth's Equator on the vernal equinox. Now, set the celestial sphere revolving from east to west. While the Earth is actually spinning from west to east, the effect is the same—stars pass overheard from east to west.

It is critical to know both your approximate latitude and the reference latitude of any star guide you're using. For example, if you were standing at the North Pole, you could not see the stars located south of the celestial equator. Travel south, however, and more of the celestial sphere below the celestial equator comes into view.

The Ecliptic
Your view of the sky depends on more than just location. What we see also depends on the time of year, and changes slightly on a daily basis. The apparent path of the sun through the year is called the ecliptic—a line that reflects Earth's orbit around the closest star.

The sun, Venus, the moon, and Jupiter in a multiple-exposure photo

The ecliptic helps explain Earth's seasonal weather, which is determined by the changing amount of time the sun spends in the sky each day, and the height it reaches between rising and setting. Earth is tilted about 23.5 degrees off of its plane of travel around the sun, but the "direction" of that tilt remains constant. At one spot in the orbit, the Northern Hemisphere tilts toward the sun and enjoys a season of longer days as the sun arcs high into the sky. These are the warm summer days. Six months later winter arrives, as the north tilts away from the sun and the geometry of daily orbit brings the sun into view for less time and at a lower angle to the horizon. The situation, of course, is reversed in the south. In the spring and fall, when the direction of tilt is "flat" to the sun, the amount of sunlight is more evenly distributed between the hemispheres.

The ecliptic determines much of what is visible to an earthbound observer. With a few bright exceptions at sunrise and sunset, the celestial objects overhead during daylight hours (and remember, the stars are always out there) are invisible to the unaided eye because of the sun's brightness. But as Earth orbits around the sun, our view constantly changes. Each nightfall brings a slightly different patch of the "sphere." Stars invisible in June, when they are washed out by the noonday sun, will be riding in the night sky six months later when Earth has moved to the other side of its orbit. Star charts are typically monthly or seasonal to take this changing view into account.

Preparing to Stargaze

The naked eye remains the most important piece of "gear" for sky-watching, combined with a vantage point as free as possible from ambient light. Getting your night vision in shape is the first step. Allow your eyes to adapt to the dark (which takes a minimum of 15 to 20 minutes) and then keep them that way. A glance at a car headlight can undo the process. Because you'll need a flashlight to read this chapter or sky charts, tape a piece of red cellophane over a regular flashlight lens or buy a red-lensed light, preferably LED. Dim red light interferes less with night vision.

Choose the darkest location you can find. At home, turn off your indoor and outdoor lights. A hill, field, or park is a better choice, as far from city light as possible. In the country, find an observation point that keeps city lights to your back. A black cloth or jacket over your head can help block out further light.

Look for a clear night; the sky is especially clear of humidity and haze following a cold front or late-afternoon storm. Developing high-pressure systems also create clear skies. It's important to avoid windy times, since wind creates turbulence that blurs your view. Be sure to dress for the chill and dampness that can creep in as the night wears on. A beach-type lounge chair offers a relaxing viewing angle, especially when settling in for a meteor show.

SKYFACT

Think binoculars aren't powerful enough to reveal much about the objects in the night sky? Then consider this: A good pair of modern binoculars are stronger than the telescopes used by 17th-century astronomer Galileo when he discovered Jupiter's moons and the phases of Venus.

THE SCIENCE OF NIGHT VISION

Before sky-watching, 20th-century astronomer Clyde Tombaugh would prepare his eyes by sitting in the dark for an hour. It can take that long for the eye to reach its peak night-vision ability. In the dark, the pupil will widen and open to let in more available light. Low- or no-light conditions allow photochemicals in the rod and cone cells to regenerate to their maximum level—a process that can take up to 45 minutes for the rods, the receptors most responsible for night vision. If you don't have an hour like Tombaugh—who would later go on to discover Pluto—even 15 or 20 minutes in the dark will help.

Equipment & Resources

As with any hobby, there is a range of gear that can take sky-watching to the nth degree, if you want to. But beginners need to acquire no more than a basic pair of binoculars to enhance their viewing. Binoculars can reveal a fantastic array of objects and details such as the larger craters on the moon, the distant planets Uranus and Neptune, and even beautiful star fields along the Milky Way. A pair of 7x50 binoculars (7 being the power of magnification and 50 the diameter in mm of the outer lens) is probably light and small enough to wear around your neck and hold steady.

A telescope is a much larger investment and should be carefully chosen. Avoid the low-end ones sold in toy stores and "big box" stores. The poor quality and results of these mass-produced instruments will only disappoint. Instead, go to a specialized store where experts can help.

Much more of what you would like to know and learn about the night sky is just a few taps away on your computer. There are many wonderful instructional and even interactive websites that offer all kinds of practical information; see a full list on pages 172-173. An example is the website for *Sky & Telescope (www.skyandtelescope.com)*, which allows you

SKY-WATCHERS & STAR PARTIES

Local museums are likely to sponsor star-watching events, or star parties, where novices and experienced amateur astronomers view the skies. A call to the astronomy department of a university or college can unlock a wide variety of resources—from a nearby planetarium to a research telescope that has hours for public viewing.

to create a custom-made naked-eye map of the sky for any location. And of course there are apps by the dozens for your mobile phone that will show you the stars and planets overhead and help you locate them. Many of these are available free or for only a small fee. But, before you turn to your gadgets, you might want to learn about the night sky in the low-tech way that humans have done for millennia. Your well-trained eyes on their own can take you a long distance into its wonders.

Sky-Watching in the City

Between the buildings blocking the horizon and the light pollution, urban sky-watching might seem a lost cause. While the subtle objects of deep space or low-to-the-horizon comets and stars might be out of reach, there is plenty left to study—particularly for the novice. The sun and moon, for example, are the two most obvious objects in the sky, and urban residents are at no special disadvantage when studying their movements. Learning the basics of these two bodies will teach you the basics of the solar system.

A warning: Special procedures and equipment, described on page 32, are needed to watch the sun safely. The intense light from that closest star, viewed through optical equipment or directly through the unaided eye, can quickly ruin your vision. But with the right precautions, sungazing can be a hobby in itself. The moon, as well—big, close, and well lit—will cut through urban glare and offers a rich amount to study: its surface topography, its phases, and its eclipses.

Beyond that, Venus reaches magnitudes as bright as -4.2, while Jupiter and Mars shine as brightly as -2.9 and -2.8, respectively—well within the limits of city viewing. Mercury can also be bright, but its orbit close to the sun limits the times when it can be spotted. Saturn, with a magnitude of roughly 0.7, can also be seen unaided even in bright conditions.

There are, in addition, 16 stars (not including the sun) with a magnitude of 1 or less. Think of them as building blocks. In the city, they will stand out in their solitude and be easier to use as star-hopping guides when you begin to acquire equipment or move to darker locations. Even in the city, the use of binoculars or a telescope can bring thousands of stars into view—if you set yourself up correctly.

PHOTOGRAPHING THE NIGHT SKY

With the right equipment, you can capture breathtaking shots of the sky—like this image, shot in the Namib Desert, showing the trails produced as stars circle the celestial pole. Photos like this can be made with a digital SLR and a long exposure or a series of photos stacked together using software. Many nightscapes can be captured with exposures no longer than 30 seconds. But extend the exposure time to minutes or hours, and you get a sky filled with star trails. Crisp night-sky images demand a rock-steady tripod. For best results, turn on any noise reduction setting in the camera and consider getting a remote release, a switch that plugs into a jack on the camera and allows the shutter to be triggered without jiggling the camera. Extra batteries are also a good idea, particularly on cool nights.

||||||||||||||||||||||||||||||||||||||||||||||||||||||||||||||||||||||||||||||||||

The Atmosphere

Any sky-watching we do is affected by the atmosphere—miles-high layers of gases, liquids, and solids that envelop the Earth. The first few layers contain most of the atmosphere's mass.

KEY FACTS

The troposphere (to 6–10 mi/10–16 km) is where clouds form and weather occurs.

+ fact: The stratosphere (10–30 mi/16–48 km) is the radiation-absorbing ozone.

+ fact: The mesosphere and thermosphere (30–210 mi/48–338 km) are where temperatures drop, then rise steeply.

+ fact: The exosphere (+210 mi/338 km) contains hydrogen and helium.

Exosphere

Thermosphere

Ionosphere

Mesosphere

Stratosphere

Troposphere

Light from the sun, stars, and other celestial objects travels great distances to reach Earth. The portion of light that reaches Earth's atmosphere must navigate the properties and activities that each layer of atmosphere contains. These cause further reductions in the light that will eventually reach the surface. The most challenging transit is through the troposphere, the layer between the surface and 6 to 10 miles (10 to 16 km) above it. This layer holds the largest concentration of atmospheric mass, including more than 97 percent of water vapor. Clouds, dust, chemical pollution, and light pollution combine there to hinder our observations of the night sky.

Nightfall

Sunset initiates nightfall, but the often spectacular colors that accompany the event reveal only part of the interactions among the Earth, the sun, the moon, and the atmosphere.

KEY FACTS

The higher the latitude, the longer twilight lasts because of the angle at which sunlight reaches Earth.

+ fact: In the far north during and near the summer solstice, twilight lasts all night long.

+ fact: The opposite phenomenon occurs in the far north's winter, when little or no sunlight arrives during the day.

As the sun sets in the west, the atmosphere disperses the sun's light like a prism. The long wavelengths of red, orange, and yellow in the spectrum reach us even after the sun has set. Meanwhile, the atmosphere scatters short-wavelength blue and violet at the opposite end of the spectrum, and almost none of these colors reach us. Look to the east and you may see a deep blue band across the horizon—the shadow of Earth, cast on the sky.

The moon and stars replace the sun. The amount of moonlight we see at night depends on the reflected light of the sun; changing amounts cause the moon's phases. Atmospheric conditions also affect the moon's brightness.

Noctilucent Clouds

In northern locations, twilight sometimes illuminates high-flying noctilucent clouds that originate far above the region where most clouds occur.

KEY FACTS

Noctilucent clouds (NLCs) were first observed in 1885, two years after the Krakatau Volcano erupted, and initially linked to the eruption.

+ fact: When Krakatau's ashes disappeared, NLCs remained.

+ fact: NLCs are also called polar meso-spheric clouds (PMCs).

+ fact: Atmospheric methane increases formation of NLCs.

Clouds typically occur in the troposphere, which ranges from 6 to 10 miles (10 to 16 km) above the Earth. Noctilucent (night-shining) clouds form at much higher altitudes in the mesosphere, which extends to 50 miles (80 km). These thin, wispy clouds appear pale white or an eerie electric blue, and they often occur above the typical orange-red layer of sunset. The arid, frigid mesosphere offers poor conditions for cloud building, but noctilucent clouds assemble from tiny ice crystals and possibly meteor smoke. They are visible in summer when the sun has dipped about 10° below the horizon and usually at altitudes above 40° N. Astronauts on the International Space Station have viewed the clouds from their unique vantage point and have taken scientifically valuable photographs of the phenomenon.

Magnetosphere

A magnetic shield sustained far beyond Earth's atmosphere by the planet's inherent magnetism, the magnetosphere deflects solar and cosmic radiation.

KEY FACTS

On the side of Earth facing the sun, solar wind compresses the magnetosphere to about 40,000 mi (64,000 km).

+ fact: Geologic studies show the magnetic field changing over time.

+ fact: The location of the north magnetic pole has wandered over arctic Canada.

+ fact: In the future, the north pole will move into Russia.

Earth's molten metal core generates a magnetic field on a cosmic scale. Magnetic current flows outward from the two poles and loops back toward the center, forming a magnetosphere that extends tens of thousands of miles into space. It deflects the steady stream of atomic particles known as the solar wind, which can knock out communications and electronics during intense geomagnetic storms. Buffeting by the solar wind causes the magnetosphere to change shape during the day. It becomes compressed on the side facing the sun, and it elongates into a long tail extending past the orbit of the moon on the opposite side. The sun and the other planets also have magnetospheres, but Earth's is the strongest among rocky planets of the inner solar system.

|||||||||||||||||||||||||||||||||||||||||||||||||||||||||||||||||||||||||||||||||||||||||||

Van Allen Belts

Consisting of two concentric zones of intense radioactivity that gird the Earth, Van Allen belts are named for James Van Allen, the physicist who discovered them in 1958.

KEY FACTS

Van Allen designed special Geiger counters that traveled on the first U.S. satellite.

+ fact: The satellite transmitted evidence of trapped cosmic and solar radiation.

+ fact: Discovery of the belts was one of the first major breakthroughs of the space age.

+ fact: Spacecraft traveling through the belts require protective shielding.

The Van Allen belts harness atomic particles that have penetrated the magnetosphere, including protons and electrons from outside the solar system and helium ions from the sun. The trapped, highly charged particles mingle with the atmosphere and travel a spiral path as they bounce back and forth between Earth's magnetic poles.

The belts can be visualized as concentric doughnuts, thickest at the Equator and weaker at the poles. They shrink and expand under different conditions. The inner belt begins approximately 600 miles (1,000 km) above Earth's surface and extends to about 3,000 miles (5,000 km). The outer belt extends from about 9,000 to 25,000 miles (14,000 to 40,000 km). In 2013, NASA's Van Allen Probes detected the temporary formation of a distant third belt.

||||||||||||||||||||||||||||||||||||||||||||||||||||||||||||||||||||||||||||||||

Auroras

A by-product of solar storms, auroras are visible mainly in polar areas. The light show is called aurora borealis in the Northern Hemisphere and aurora australis in the Southern Hemisphere.

KEY FACTS

Auroras shimmer because solar storms stretch the magneto-sphere, and it bounces back into place.

+ fact: Northern peoples developed myths to explain them.

+ fact: Southern Hemisphere auroras appear mostly over uninhabited land.

+ fact: At the peak of a solar-storm cycle in 1989, auroras were seen as far south as the Caribbean.

During violent solar events, such as mass coronal ejections and solar flares, the discharge of highly charged particles from the sun increases astronomically, shooting out billions of tons of matter at speeds as fast as 2 million miles (3 million km) an hour. This barrage breaches the magnetosphere and Van Allen belts, reaching perhaps as close as 50 miles (80 km) to the Earth's surface.

The abundant solar material energizes molecules of oxygen and nitrogen in the atmosphere, causing dazzling curtains of green and red and pink light that intensify in proportion to the intensity of the solar storms in their 11-year cycle.

||||||||||||||||||||||||||||||||||||||||||||||||||||||||||||||||||||||||||||||||

Stars

From their earliest formation about 13 billion years ago, stars have been a universal sky feature. A few thousand can be seen with the naked eye on a clear, moonless night.

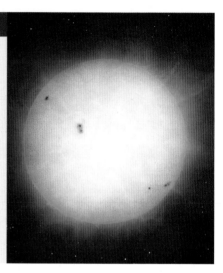

KEY FACTS

The sun is a third-generation star with elements from two previous star explosions.

+ fact: The sunlight currently shining on Earth was generated in the star's core 50 million years ago.

+ fact: Sunlight takes 8.3 seconds to reach Earth.

+ fact: A neutron star fragment the size of a sugar cube would weigh about 100 million tons on Earth.

Elementally simple—composed almost entirely of hydrogen and helium gases—stars have not yet revealed the precise details of their formation. In basic outline, gravity pulls inward dense patches of gases and dust, causing pressure and heat to build at the center of the ball of gas and triggering nuclear fusion. The energy created, which we see as starlight, pushes outward and is balanced by the pull of gravity, establishing equilibrium. Stars are born in nebulae, large gas clouds found throughout the Milky Way. They also come into being from the supernova explosions of older stars that create raw material for new stars. The nuclear reactions within stars form the heavy elements carbon, oxygen, and nitrogen—the essential building blocks of life as we currently understand it.

‖‖

Types of Stars

Astronomers classify stars by the characteristics of size, temperature, and color. They use hydrogen spectral analysis of stars' hydrogen emission lines to make their determinations.

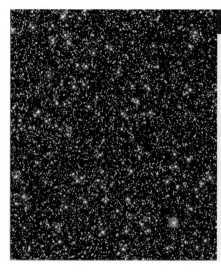

KEY FACTS

Stars are classified by a letter system that runs O, B, A, F, G, K, M (largest/hottest to smallest/coolest).

+ fact: A mnemonic to remember the order is "Oh boy, a flying giraffe kicked me!"

+ fact: A Hertzsprung-Russell diagram shows the relationship between color, temperature, and luminosity of stars.

+ fact: The sun is a modest G star.

A star's size is determined by its mass and not by a linear measurement, such as diameter. Mass regulates the other properties of a star, including temperature, color, and longevity. Smaller, cooler stars burn fuel at a slower pace and live longer than larger, fuel-guzzling hot stars. Red dwarfs are the most common smaller stars and have potential life spans of tens of billions of years.

Massive blue giants are the hottest and may burn out in a million years or so, relatively young for a star. The supergiant is the largest star; Betelgeuse in Orion is a red supergiant, with a mass about 1,000 times bigger than the sun. Despite its large size, a red supergiant is a relatively cool star. It burns approximately ten times cooler than hotter and smaller blue O stars.

|||||||||||||||||||||||||||||||||||||||||||||||||||||||||||||||||||||||||||||||||||||

Star Brightness

We talk about star brightness in terms of magnitude, a number that often represents how bright the star is from our vantage point on Earth.

KEY FACTS

Sirius, a blue-white binary star in the constellation Canis Major, is the brightest star in the Northern Hemisphere.

+ fact: Second brightest is Arcturus, an orange giant in Boötes.

+ fact: Fifth is Vega in Lyra. (Third and fourth are visible only in the southern sky.)

+ fact: Sixth is Capella in Auriga.

I n the second century B.C., the Greek astronomer Hipparchus created a catalog that ranked the brightest stars as of the first magnitude, less bright stars as the second magnitude, and so on. Adopted by Ptolemy and refined in later centuries, the system became a logarithmic scale. As observation methods improved and fainter stars came into view, the scale was extended at the higher (dimmer) end. The need also arose to accommodate differences among first-magnitude stars. With nowhere to go but down, negative numbers were added. Magnitude is now calculated in many ways, including absolute magnitude, which is brightness if all stars were placed at the same distance from Earth. Under average night sky conditions, the naked eye can see stars to magnitude +6.

Star Death

The state of equilibrium that keeps a star together is a temporary one. As a star's core consumes its fuel, it sets in motion the scenario of its death.

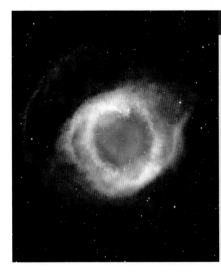

KEY FACTS

A spent white dwarf often hooks up with a vibrant star and travels with it.

+ fact: Brightest star Sirius travels with the Pup, not a young star, but an old white dwarf.

+ fact: Neutron stars, also known as pulsars, emit beaconlike pulses of radiation.

+ fact: A supernova explosion creates large fields of new material for star formation.

As a star burns up its fuel, its nuclear core begins to shut down. Gravity can no longer be held at bay, and the star contracts. This raises its temperature and triggers renewed fusion, allowing the star to expand. Smaller stars will become pulsing red giants. Eventually, they shed outer layers and become dead stars known as white dwarfs. Larger stars morph into supergiants, with contracting cores that may reach 1,000,000,000°F. More intense gravity in the core leads to an implosion followed by an explosion—a supernova—that blows apart the outer layers. Next follows either a neutron star or a total collapse into a black hole.

Sun

Our sun came into being the way any star does—in a cauldron of condensed gas and fused atoms. This life-sustaining star sits at the center of our solar system.

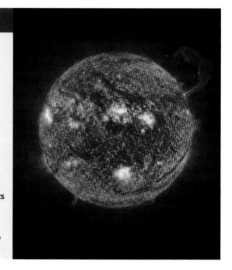

KEY FACTS

For a star, the sun is of middling size and heat—10,000°F (5500°C) at surface.

+ fact: Sun's core temperature: about 27,000,000°F (14,999,820°C)

+ fact: Composition of sun's outer layer: 74 percent hydrogen, 25 percent helium, 1 percent heavy elements

+ fact: The sun provides 1,400 watts of energy to each square yard of Earth.

Despite its seemingly extreme statistics, the sun is the perfect star for our planet. One of about 200 billion stars in the Milky Way, it resides 93 million miles (150 million km) from Earth. Its dimensions, dynamics, and distance provide just what we need in terms of energy and light to support life without overwhelming the planet with violent radiation. The sun is classified as a yellow dwarf star and lies in the middle of the classification scheme for color and temperature. Earth's life-supporting relationship with the sun is jeopardized by depletion of the upper ozone layer in the atmosphere, which allows more ultraviolet light to reach the planet. As with any star, the sun has an expiration date, estimated at about five billion years from now. It is currently about halfway through its lifetime.

II

Sun Anatomy

The sun is a roiling, complicated body of violent extremes, with its own weather systems and a disturbing hint of unpredictability.

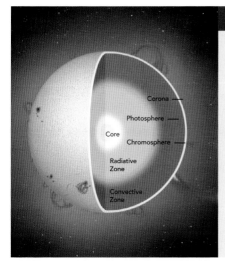

KEY FACTS

The sun's equatorial region rotates in 25 days, but its polar regions take 34 days to complete rotation.

+ fact: In the core, about half the sun's mass is packed into 7 percent of its volume.

+ fact: Nuclear fusion in the core generates 400 trillion watts of energy a second.

+ fact: Holes in the corona allow the flow of a steady stream of solar wind.

The sun has a dense, high-energy core, where fusion takes place and hydrogen is converted to helium; emitted photons begin a journey to the surface that can last millions of years. Three-fourths of the way there, they reach the convective zone, which continues to draw gas and heat outward. They then encounter the 300-mile-thick (483 km) photosphere, which gives the sun the appearance of a solid with gas bubbling at the surface. The chromosphere is a pinkish layer that harbors solar flares. Beyond that lies the corona, a billowing megahot halo with temperatures reaching 2,000,000°F (1,111,093°C) that extends millions of miles. Holes in the corona, caused by the sun's magnetic field, allow a steady stream of particles known as solar wind to break free.

Sunspots

Sunspots peak and ebb in an 11-year cycle that coincides with other solar activity. Chinese astronomers made note of sunspots in 28 B.C.

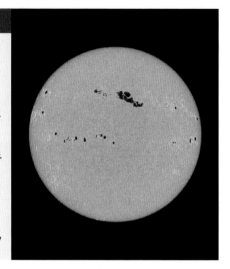

Disrupted currents in the sun's magnetic field can slow the flow of solar material to the sun's surface, causing it to cool and darken in color. This produces sunspots, dark, visible splotches on the sun's photosphere. The center of a sunspot, depressed slightly below the surrounding gases, can measure some 3600°F (1980°C) cooler than the sun's typical surface temperature of 10,000°F (5500°C). Within a cycle, sunspot activity usually starts in the regions of latitudes 30° N and 30° S and then moves near the sun's equator. The 11-year cycle is part of a 22-year period in which the magnetic fields in the sun's upper and lower hemispheres switch polarities. NASA's Solar Dynamics Observatory is designed to study the variations in solar activity that affect Earth and near-Earth space.

‖‖

Solar Flares

One thing leads to another: Solar flares, immense eruptions of high-energy radiation from the sun's photosphere, are associated with sunspots in ways not completely understood.

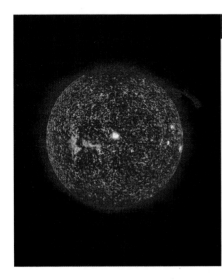

KEY FACTS

One solar flare can unleash more energy than all the atomic bombs ever detonated.

+ fact: The first solar flare was recorded in astronomical literature in 1859.

+ fact: In 1989, intense solar activity knocked out much of Quebec's electrical grid.

+ fact: NASA and other science organizations predict solar storms.

When magnetic energy that is pent up as a result of sunspot activity breaks free, it releases intense bursts of radiation across the entire electromagnetic spectrum, from radio waves to gamma waves. Solar flares extend into the sun's corona and take the form of arcs of highly charged superheated material known as coronal loops. The solar storms produced by flares sabotage electronics, communications, and electrical grids as they unleash their disruptive radiation. The most troublesome kinds of solar flares are known as X-class flares, which produce prolonged radiation storms in the upper atmosphere.

Sun's Path

For centuries, we have known that the Earth orbits the sun, but to conveniently describe all kinds of solar and celestial phenomena, we talk of the sun's apparent path across the sky.

KEY FACTS

Earth rotates from west to east, making the sun's path appear to be from east to west.

+ fact: Perihelion is the point at which a planet or other sky object in orbit is closest to the sun.

+ fact: Aphelion is the point at which a planet or other sky object is farthest from the sun.

We on Earth cannot track in a useful way our planet's daily rotation or yearly orbit without reference to the sun. The sun's apparent daily journey across the sky—and the changes to its route over the course of a year—have long formed the means by which we tell time, mark the seasons, and plan annual celebrations.

The path of the sun across the sky is known as the ecliptic, which is also the term given to the Earth's route around the sun. Most sky charts indicate the sun's ecliptic on the night sky as a useful reference for locating constellations and planets and describing their whereabouts. The constellations that form the zodiac follow the path of the sun's ecliptic over the course of a year, the basis of the 12 astrological signs.

Sun & Seasons

Earth and sun are oriented to each other and interact in a way that produces predictable periods of difference in sunlight and temperature throughout the year.

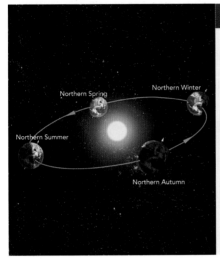

KEY FACTS

During the equinoxes, the division of day and night is about the same around the globe.

+ fact: Summer at the North Pole means nearly 24 hours of daylight.

+ fact: The Earth's axis moves in position a little more than a half degree a century.

+ fact: The change of direction in the Earth's axis is known as precession.

As Earth orbits the sun, its rotational axis is tilted 23.5 degrees off the perpendicular. Because of this, the sun's direct light will hit different parts of the Earth at different times of the year, causing the seasons. Astronomical convention marks the transitions in two ways: as equinoxes in the spring and autumn, when the sun at noon shines directly on the Equator; and as solstices in winter and summer, when the sun shines on Earth as it is tilted away from or toward the sun.

The Northern and Southern Hemispheres experience seasons at opposite times of the year, and equatorial areas experience little seasonal variation at all. Seasonal differences in sunlight and temperature affect many of the life processes on Earth.

Viewing the Sun

Many of us quickly reach for our sunglasses when we hit the bright sun—a good instinct. Viewing the sun directly with the naked eye can cause permanent damage.

Here is a mantra to repeat over and over for sky-watchers young and old: Never, ever look directly at or point your binoculars or telescope toward the sun—not even for a split second. That said, there are safe ways to observe the sun. Welders' goggles rated 14 or higher are one option, as are approved solar filters and "eclipse glasses." These items can be found online. It is also easy to make a pinhole projector by piercing a card with a small hole and placing a whole card below it. Position the top card toward the sun, without looking. A binocular lens can be used the same way; leave the caps on the unused lens. Using exposed film, medical x-rays, or smoked glass as filters is dangerous. Eye damage may be painless, as it occurs because it is caused by invisible infrared rays.

III

Solar System

We tend to talk of *the* solar system—ours—but there are many of them. Our sun's gravity draws planets and other objects into orbit around it.

KEY FACTS

More than 99 percent of the solar system's mass is contained within the sun.

+ fact: Primitive meteorites provide a measure of the age of the solar system: about 4.5 billion years old.

+ fact: The solar system orbits the center of the Milky Way every 220 million years.

+ fact: The word "planet" means "wanderer."

The sun holds sway over a diverse array of objects ranging from planets and dwarf planets to moons and asteroids—and even to dust. Its influence spreads from its own surface to the distant Oort cloud that encircles the solar system. The planets under the sun's control—including Earth, of course—spread across a distance ranging from Mercury, on average about 36 million miles (58 million km) from the sun, to Neptune, with an orbit averaging about 2.8 billion miles (4.5 billion km) from the sun. Pluto and other dwarf planets range beyond that. Of course, our solar system is only one of many that have formed or at this moment are forming around distant stars. Ongoing discoveries attest that one in six stars has an Earth-size planet in tight orbit.

Moon

The moon is a stargazer's nearly constant companion, visible to the naked eye even in competition with extreme light pollution. Binoculars bring out details of its features.

KEY FACTS

Made of lighter elements than Earth, the moon has $1/4$ the diameter of the Earth, contains about $1/80$ the mass, and has about $1/6$ the gravity.

+ fact: Moon's diameter is 2,159 mi (3,465 km).

+ fact: Distance from Earth is 238,855 mi (384,400 km).

+ fact: Orbital period is 27.3 days; cycle of moon phases is 29.5 days.

Moons are natural satellites that orbit celestial bodies such as planets and asteroids, and exercise their own gravitational pull. Most planets in our solar system have moons, usually multiples. Earth's solitary moon likely formed from debris ejected when an object about the size of Mars struck the young Earth, about 4.5 billion years ago. Evidence from moon missions shows a similar composition of their respective rocks. Characteristics of their orbits also support this theory. Over time, the moon acquired its craters and seas from bombardments by space debris and the eruption of molten rock from the interior. The "man in the moon" is an impression that the unaided eye sees of contrasting lunar surface features. Some people see a woman, a hare, or a toad.

Moon in Motion

The motion of the moon shares complex interactions with that of the Earth. The planet and its satellite regulate and influence each other in various ways.

KEY FACTS

Friction from Earth-moon interaction causes the Earth to slow slightly.

+ fact: The slowdown is offset by a slight pulling away of the moon—about 1.5 in (3.8 cm) a year.

+ fact: The moon's orbital period is 27.3 days.

+ fact: The Earth's rotation accounts for much of the moon's visible motion.

The moon spins on its axis in the same manner as the Earth. But the moon's rotation takes the same amount of time as its orbit around the Earth—27.3 Earth days. This is because the Earth's gravity produces a bulge, or "land tide," in the moon that slows it down. As a result of this synchronization of rotation and orbit, the moon always presents the same side to the Earth. The moon still exerts its own influence: For example, it creates a gravitational tug responsible for the tides on Earth. The moon's gravity requires spacecraft engineers to take it into consideration when calculating trajectories for lunar landings. World cultures have attached other influences to the moon, such as influencing the course of love or initiating the legendary transformation from human to werewolf.

Moon Phases

The moon "borrows" light from the sun to create its shine. The phases, or appearance, of the moon at different times of its orbit reflect changes in available sunlight.

KEY FACTS

The moon takes 29.5 days to complete the cycle of its phases.

+ fact: Moonrise occurs 12 degrees farther to the east each night.

+ fact: A full moon rises with the setting sun and sets at sunrise the next day.

+ fact: View crescent moons best in the evening in spring and just before sunrise in autumn.

Observers on Earth cannot see the moon during its phase known as new moon. It occurs when the moon lies between the Earth and the sun and is fully backlit. About two weeks later, the Earth lies between the sun and moon, and we see the latter's maximum appearance—a full moon. In the weeks leading to the full moon, the moon comes into view in gradual increments: a waxing crescent, crescent, first quarter, and then waxing gibbous. (Gibbous is from the Latin for "hump.")

In the weeks after the full moon, its light starts to diminish, and it becomes a waning gibbous, last quarter, crescent, and then waning crescent. Worldwide, many agricultural tasks as well as religious observations are linked to the phases of the moon.

Moon's Near Side

Collisions with meteors and comets, combined with volcanic activity from the moon's interior, produced the landscape of ridges and craters we see on the near side of the moon.

KEY FACTS

Each part of the moon's surface is visible for about two weeks in a phase cycle.

+ fact: Galileo observed the moon's surface through his telescope in the 1600s.

+ fact: First and last quarter phases show greater detail on the moon's surface.

+ fact: The last major crater to form was the Tycho crater about 109 million years ago.

Earth's moon began in violence that continued for the first half billion or so years. Debris bombarded the surface, creating large, wide, and shallow areas that later filled with dark lava pouring from the moon's core. These areas are called "maria" (MAH-ree-uh), the Latin word for "seas." The most famous is the Mare Tranquillitatis, the Sea of Tranquillity, site of the first moonwalk by Neil Armstrong in 1969. Smaller debris gouged smaller, deeper holes known as craters. These appear much lighter on the moon's surface than the dark maria. Over time, the bombardment decreased dramatically. NASA's Astronomy Picture of the Day (APOD), available at *apod.nasa.gov/apod/astropix .html*, often features amazing images of the moon, including some interactive ones.

Moon's Near Side

Moon Features

The moon's synchronous orbit with Earth—spinning on its axis in the same time it takes to move around the planet—means only one side is visible to land-based observers. The visible side includes lighter colored, cratered highlands, where collisions with meteors and comets have left dozens of broad, deep holes. The large, dark maria are places where molten rock has oozed out and filled large impact basins, leaving a smoother surface. Seas and craters are labeled on the accompanying chart for your reference.

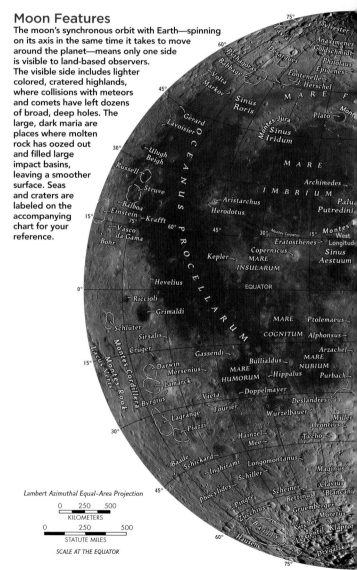

Lambert Azimuthal Equal-Area Projection

KILOMETERS
0 250 500

STATUTE MILES
0 250 500

SCALE AT THE EQUATOR

Gazing at the Moon

Each part of the moon gets two weeks of light as it moves through its phases. Details are best studied about a week after and before a full moon, when sunlight strikes the surface at an angle, and along the line known as the terminator, which divides the sunlit portion of the moon from the dark part. Crescent moons are best spotted in the evening in the spring and just before sunrise in the autumn.

North Pole

75°
60°
45°
30°
15°
0°
15°
30°
45°
60°
75°

Byrd
Challis
De Sitter
Petermann
Baillaud
Cusanus
Arnold
Strabos
Belkovich
MARE HUMBOLDTIANUM

Bartow
Neison
Kane
W. Bond
De La Rue
Gärtner
Endymion
Mercurius
Zeno

G O R I S
Aristoteles
Atlas
Riemann
Messala
Gauss

Eudoxus
Hercules
Lacus Mortis
Franklin
Berosus
Hahn

Cassini
Lacus Somniorum
Geminus
Cleomedes
Rayleigh
Plutarch

Alexander
Posidonius
Montes Taurus
MARE ANGUIS
Hubble

Aristillus
MARE
SERENITATIS
Goddard
MARE MARGINIS

Montes Caucasus
MARE CRISIUM

Montes Haemus
Palus Somni
Condorcet
Jansky

East longitude
MARE VAPORUM
Julius Caesar
MARE
MARE TRANQUILLITATIS
Firmicus
MARE UNDARUM
Neper

MARE SPUMANS
MARE SMYTHII

Hipparchus
Sinus Asperitatis
MARE FECUNDITATIS
Kästner
Gilbert

Theophilus
Gutenberg
Langrenus
Kiess

Albategnius
Cyrillus
Montes Pyrenaeus
La Pérouse
Ansgarius
Lamé
Behaim

Parrot
Catharina
MARE NECTARIS
Vendelinus
Hecataeus

Sacrobosco
Beaumont
Santbech
Wrottesley
Phillips

Blanchinus
Apianus
Rupes Altai
Fracastorius
Petavius
Humboldt

Werner
Piccolomini
Snellius
Legendre

Aliacensis
Zagut
Stevinus
Barnard
Adams

Walther
Riccius
Rheita
Furnerius
Harlan
Abel

Stöfler
Metius
Fabricius
Oken
Gum

Licetus
Maurolycus
Janssen
Hamilton

Barocius
Clairaut
Pitiscus
Watt
Lyot

Lilius
Vlacq
Rosenberger
Biela
Hanno

Zach
Nearch
Pontécoulant
Hagecius
Gill

Curtius
Helmholtz
Neumayer

Boussingault
Demonax
Scott
Amundsen

South Pole

Moon's Far Side

Although it is not a visual presence in the night sky, understanding the moon's far side rounds out our appreciation of Earth's satellite.

KEY FACTS

Many far-side features have names that reflect Russian space exploration, such as the Moscow Sea.

+ fact: The far side experienced fierce asteroid bombard-ment, causing craters.

+ fact: Bombardment on the far side caused volcanic activity on the near side.

+ fact: The far side has a thicker crust with fewer seas than the near side.

The moon's far side is often referred to as its "dark side," but it is by no means dark on the far side; we just cannot see the light because, due to its synchronous orbit with Earth, the moon presents only its near side to our view. The far side experiences phases just as the near side does; there is a time each month when it is in full sunlight and a time when it is in full shadow. The far side of the moon was first mapped from space-based observations by the Russians in 1959, with pictures the spacecraft Luna 3 took. After that, three decades of probes and space missions created detailed maps. The quest continues to provide even more detailed and scientifically advanced topographical images of the moon's far side with the Lunar Reconnaissance Orbiter, launched by NASA in 2009.

Moon Exploration

On July 20, 1969, men on the moon—the astronauts of Apollo 11—
met the man in the moon, so to speak. Remote reconnaissance
and robotic landings preceded the historic event.

KEY FACTS

Scientists believe that
most of the 6 U.S.
flags planted on the
moon still stand.

+ fact: In the 1960s,
orbiters photographed
99 percent of the
moon's surface.

+ fact: The last Apollo
mission, in December
1972, collected more
than 250 lb (113 kg)
of moon rocks.

+ fact: India, China,
and Japan are devel-
oping lunar explora-
tion projects.

The manned Apollo missions with their moon landings
and space buggy rides were preceded by remote data
gathering. The Russians sent the first probes to check out
the moon in 1959: Luna 1 flew by it, Luna 2 crash-landed,
Luna 3 took photographs of the far side; and in 1966, Luna
9 made a soft landing. NASA followed with several series
of lunar probes, including orbiters and surveyors that
paved the way for six manned Apollo
missions and their 12 astronauts.
NASA maintains a Lunar Reconnaissance Orbiter that acquires
sophisticated data, and the space
agency is committed to returning
humans to the moon.

Solar Eclipse

The sun and moon collaborate to produce three kinds of solar eclipses, obscurations of the sun by the moon. These are called annular, partial, and total.

KEY FACTS

Viewing solar eclipses requires eye protection.

+ fact: Many cultures feared solar eclipses, believing they portended bad fortune.

+ fact: The slender trail of a total solar eclipse varies year to year in an 18-year cycle.

+ fact: The next path of a total solar eclipse to cross the U.S. will occur in 2017.

umbra

penumbra

An annular eclipse occurs when the moon is too far from the sun to cover it completely. Instead, a ring, or annulus, of the sun's bright edge is visible. A partial eclipse happens when only part of the sun is obscured if viewed from within the moon's penumbra, the outer region of its shadow. Annular and partial eclipses only dim the sun. During totality of a total solar eclipse, the moon's disk completely covers the sun for those on Earth within the moon's umbra, or central shadow. Daylight fades to dark twilight, and stars are visible during totality. Sky-watchers can see the sun's corona feathering out around the edges of the moon's shadow.

Lunar Eclipse

Like the sun, the moon experiences three kinds of eclipses: penumbral, partial, and total. Unlike solar eclipses, they can be viewed with the naked eye.

umbra

penumbra

KEY FACTS

Columbus used his knowledge of an impending lunar eclipse to gain the cooperation of Arawak Indians, who were amazed by his ability to predict it.

+ fact: Ancient sky-watchers tracked lunar eclipses and kept tables of their occurrences.

+ fact: Use binoculars or a telescope with a low-power eyepiece to see the whole moon.

If the moon passes through the outer region of the Earth's shadow, the penumbra, it causes a penumbral lunar eclipse. This will dim the moon a bit. If some of the moon passes through the Earth's dark central shadow, the umbra, a partial eclipse will occur. When the moon moves completely into the Earth's umbra, a total eclipse takes place. Even then, the moon does not become completely dark. Sunlight bends around Earth's surface and shines dimly on the moon. The refracted light casts a reddish or orange glow. If Earth's atmosphere is unusually dusty, the moon might be barely visible for a while.

||||||||||||||||||||||||||||||||||||||||||||||||||||||||||||||||||||||||||||||||||

Satellites

The U.S.S.R. launched the first artificial satellite, Sputnik 1, into orbit on October 4, 1957. Since then, thousands more have been launched for scientific and commercial purposes.

KEY FACTS

Sputnik 2 carried a dog named Laika. Her flight proved that an animal could survive launch and weightlessness, although she perished shortly after.

+ fact: In 1959, U.S. satellite Explorer 6 photographed Earth for the first time.

+ fact: In 1962, Telstar, the first communications satellite, was launched.

A satellite is any object that orbits a planet; Earth's moon is a natural satellite. Artificial satellites were created and sent into space for purposes of data gathering and signal transmission. The launch of Sputnik 1 in 1957 ushered in the space age. The United States followed on January 31, 1958, with the launch of American Explorer 1—and the space race between the U.S.S.R. and the U.S. was under way. In the meantime, satellite launchings have become commonplace. There are roughly 6,000 satellites in orbit now, not all of them functional, with specialized missions such as communications, weather, navigation, and astronomy.

Highly Elliptical Orbit

Geosynchronous Orbit

Low Earth Orbit

Semi-synchronous Orbit

Cassini-Huygens Spacecraft

A joint venture of NASA, the European Space Agency, and the
Italian Space Agency, the Cassini-Huygens spacecraft launched in
October 1997 to begin the long journey to Saturn.

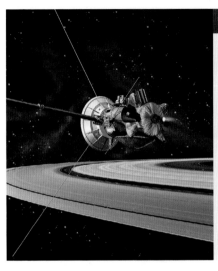

KEY FACTS

The Huygens probe
was the first craft to
land on a body in the
outer solar system.

+ fact: Seventeen
countries collaborated
on the Cassini-Huygens
mission.

+ fact: It takes more
than an hour for radio
signals from Cassini to
reach Earth.

+ fact: The Cassini
mission has been
extended to 2017.

Saturn, with its rings suggestive of the sun's encircling
gases and its many moons representing diverse aspects
of planet formation, was an ideal target for the Cassini-
Huygens mission. The hyphenated name reflects the
craft's two components: the plutonium-powered Cassini
orbiter and the wok-shaped Huygens probe, named for
Italian-French and Dutch astronomers, respectively. Cassini-
Huygens took nearly seven years to reach Saturn. Huygens
parachuted to a soft landing on Saturn's moon Titan in
January 2005. The duo has made pathbreaking discoveries,
including strange weather patterns, new moons, methane
lakes on Titan, and ice geysers on the moon Enceladus.
Cassini is on a second mission extension until 2017, when
the Saturn summer solstice can be observed.

||||||||||||||||||||||||||||||||||||||||||||||||||||||||||||||||||||||||||||

Chandra X-ray Observatory

The Chandra X-ray Observatory, named for Indian-American astrophysicist Subrahmanyan Chandrasekhar, provides a view of the universe not possible from Earth's surface.

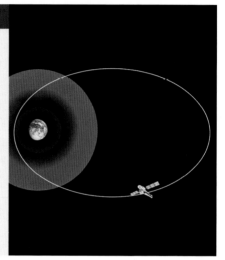

KEY FACTS

Chandra can see the x-rays from particles at the edge of black holes.

+ fact: Its orbit takes it 200 times higher than the Hubble Space Telescope.

+ fact: Its resolving power compares to reading a stop sign from 12 mi (19 km) away.

+ fact: Chandra needs only the wattage of a hair dryer to run the spacecraft.

Launched in July 1999 by the space shuttle Columbia, NASA's Chandra X-ray Observatory is a telescope that focuses on the high-energy objects in space that radiate energy in x-ray wavelengths, such as supernovae, binary stars, and black holes. The world's most powerful x-ray telescope, it can detect x-ray sources only one-twentieth as bright as earlier telescopes could. To do so, it orbits above Earth's atmosphere, up to an altitude of 86,500 miles (139,000 km). Chandra, a NASA project that is hosted by the Harvard-Smithsonian Center for Astrophysics, has thrilled astronomers with its data, which it shares with scientists around the world.

||||||||||||||||||||||||||||||||||||||||||||||||||||||||||||||||||||||||||||

Hubble Space Telescope

Launched in 1990, the Hubble Space Telescope is the astronomical equivalent of the greatest thing since sliced bread. It looks at the universe in mind-boggling detail.

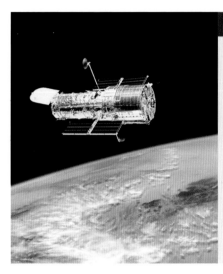

KEY FACTS

The Hubble travels at 17,000 mph (28,000 kph) and orbits 353 mi (569 km) above the Earth.

+ fact: Every year, the Hubble makes more than 20,000 observations.

+ fact: Space shuttle missions have upgraded and repaired equipment, including cameras on the Hubble.

Its ability to view far-flung reaches of the universe without atmospheric interference is the hallmark of the Hubble Space Telescope. Roughly the size of a large bus, the Hubble carries a primary mirror 7.9 feet (2.4 m) in diameter and an array of instruments, including cameras, spectrographs, and fine-guidance sensors used to aim the telescope. In the more than two decades since its launch, a relentless stream of images from the Hubble has helped determine the age of the universe, informed us about quasars, and enlightened us about dark energy. Its successor, the infrared James Webb Space Telescope, is in development. The Webb, planned to launch in 2018, will have a mirror nearly three times the size of the Hubble's and a sun shield the size of a tennis court.

|||||||||||||||||||||||||||||||||||||||||||||||||||||||||||||||||||||||||||||||||||||||

Voyagers 1 & 2

Launched more than 35 years ago, NASA's Voyagers 1 and 2 are now headed out of the solar system, prepared for encounters with intelligent life.

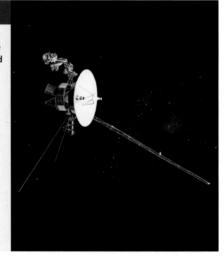

KEY FACTS

The probes' mission originally was to study Jupiter and Saturn and their moons.

+ fact: Early success led to an expansion and continuation of the Voyager mission, with flybys of all outer planets.

+ fact: Voyager 1 is the farthest artificial object in existence—119 times farther from the sun than the Earth is.

The deep-space probes Voyager 1 and 2 were launched in 1977 to study the outer solar system and, eventually, interstellar space. These nuclear-powered craft are headed out of the solar system on two separate trajectories, with Voyager 1 well ahead. Should either encounter intelligent life, each carries a gold-plated 12-inch (30 cm) phonograph record—very "old school," but the go-to durable format at the time. Drawings on the cover demonstrate how to play the record, and they show a map of our solar system relative to 14 beaconlike pulsars. Music included ranges from Bach to Javanese gamelan to Navajo chant to Chuck Berry's "Johnny B. Goode." The record also contains spoken greetings in 55 world languages and natural sounds, such as thunder, birds, and whales.

Space Junk

What goes up into space may come down, but it often stays up and accumulates with other outmoded hardware to become orbiting space junk.

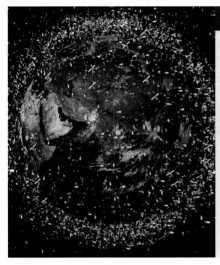

KEY FACTS

The now nonfunctional Vanguard 1 satellite, launched in 1958, is expected to orbit another 150 years.

+ fact: Space junk endangers weather, communications, and GPS satellites.

+ fact: When space junk falls out of orbit, it can cause potential damage on Earth.

+ fact: In 1978, a Soviet satellite spewed radioactive debris into the Arctic.

With the exception of manned spacecraft, most satellites and other equipment sent into space do not make the return journey. Even when no longer functional, these objects remain in orbit to become part of a growing array of space junk. Some 600,000 items ranging from rocket boosters to explosion debris to actual space station garbage float out there, creating hazards to functioning spacecraft and satellites from time to time. At most peril is the International Space Station, or ISS, and its inhabitants. In 2001, the ISS had to be nudged into a slightly higher orbit by the space shuttle *Endeavour* to avoid a collision with the upper stage of a spent Soviet rocket.

‖‖

Space Station

Sky-watchers are often treated to the sight of the International Space Station, or ISS, in the night sky. NASA's website and its SkyWatch 2.0 applet provide information on its whereabouts.

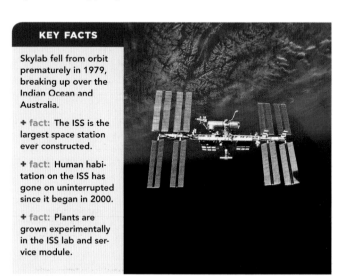

KEY FACTS

Skylab fell from orbit prematurely in 1979, breaking up over the Indian Ocean and Australia.

+ fact: The ISS is the largest space station ever constructed.

+ fact: Human habitation on the ISS has gone on uninterrupted since it began in 2000.

+ fact: Plants are grown experimentally in the ISS lab and service module.

Prolonged sojourns in orbit require a space station. Both the Soviet Union (and later Russia) and the United States committed to this endeavor, beginning with Salyut 1 in 1971. The Soviet Salyut series was succeeded by the Mir space station, which operated from 1986 through 2006. The U.S. launched Skylab in 1973, and it was manned through 1974. Today, the International Space Station is the sole station in orbit, and it serves the exploration and research needs of the U.S., Russia, Japan, Canada, and 11 European countries. The ISS received its first crew in November 2000, and is expected to remain in operation until at least 2020. For the One Year Mission from 2015–2016, two astronauts resided on the ISS for 340 days to examine human physiological responses to prolonged space travel.

UFOs

Few topics are as controversial as unidentified flying objects—
UFOs. Despite many unexplained sightings, UFOs often turn out
to be ordinary phenomena or objects.

KEY FACTS

The term "flying sau-
cer" was coined by a
pilot in 1947.

+ fact: Over the
decades, NASA astro-
nauts have reported
a number of unex-
plained phenomena.

+ fact: In 2012,
Mars rover Curiosity
appeared to capture
a UFO image that
turned out to be
dust kicked up by the
impact of its sky crane
with the surface.

Beginning sky-watchers see a lot of objects they cannot
identify, but that doesn't make those objects UFOs. The
watchers soon learn to sort out planes, larger satellites, and
sometimes even the International Space Station from relatively stationary objects in the sky. Otherwise, UFOs in general can usually be accounted for as meteors, clouds, birds,
weather balloons, atmospheric phenomena such as sun dogs,
or military aircraft. The often secret nature of military aircraft
leaves many sightings unexplained. A large number of unre-
solved UFO sightings may well be unfamiliar atmospheric
phenomena with unusual physical properties. The SETI Insti-
tute searches for signs of extraterrestrial life using radio and
optical telescopes. Many prominent scientific institutions, as
well as NASA, have collaborated on the project.

||||||||||||||||||||||||||||||||||||||||||||||||||||||||||||||||||||||||||||

Comets

Dirty balls of frozen gases and dust, comets are leftovers from the formation of the solar system. Nudged out of their original orbits, they streak through the inner solar system.

KEY FACTS

Comets typically are large enough to survive hundreds of trips through the inner solar system.

+ fact: British astronomer Edmond Halley never saw the comet named for him.

+ fact: The most recurrent comet is Encke, cycling past Earth every 3.3 years.

+ fact: A stream of dust shed by a comet's nucleus forms a second tail.

When an icy ball of gases and dust enters a new orbit and approaches the sun, the warming object forms a glowing head—the coma—and a flow of charged ions is shaped by solar wind into a glowing tail. Long-period comets come from the Oort cloud, a belt of icy debris that encircles the solar system. Short-period comets, with orbits easy to predict, originate in the Kuiper belt beyond Neptune or in a far-flung area known as the scattered disk.

Comets can be viewed with the naked eye and, of course, through optical equipment. To tell a comet from a star, check a star chart to see if it's there, look for a tail, and look for movement over time. The NASA website offers an Asteroid and Comet Watch that provides up-to-date information.

Meteors

Called "shooting stars," meteors are bits of material, often left behind by a comet, that ignite and burn up as they pass into Earth's atmosphere.

KEY FACTS

More than 80 meteorites found are pieces from the moon, and more than 60 are fragments of Mars.

+ fact: Comets under the pull of Earth's gravity reach speeds of 20,000–160,000 mph (32,000–257,000 kph).

+ fact: Atmospheric friction heats meteors to 2000°F (1093°C).

+ fact: Most meteors vaporize before they get within 50 mi (80 km) of Earth.

Meteors originate as meteoroids, bits of interstellar dust and debris. The particles may be very small or quite large, approaching asteroid size, although the difference between the two is somewhat arbitrary. Larger meteoroids may be broken-off pieces of asteroids, planets, or our moon. Many smaller meteoroids represent burned-off ashes from comets, which are left in massive trails throughout the solar system. As Earth travels in its orbit, it encounters perhaps a half billion meteoroids each year. Objects entering the atmosphere are known as meteors, and those that reach the Earth are called meteorites. On a given night, there may be several visible shooting stars each hour. Up to 10,000 tons of meteoric material, much of it dust size, fall on Earth daily.

Meteor Showers

About 30 times a year, Earth passes through a dense dust trail, typically left by a comet, giving rise to a meteor shower or a rare, spectacular meteor storm.

KEY FACTS

The International Meteor Organization depends on information from amateurs to expand its database.

+ fact: Meteor viewing is affected by atmospheric conditions and light pollution.

+ fact: Halley's comet left a dust trail that produces two meteor showers each year.

Meteor showers, unlike the sporadic meteors, occur in steady streams at predictable times from specific locations in the sky known as radiants. Radiants typically are identified with constellations. The Leonid meteor shower, for example, originates in Leo. Meteor showers may produce several dozen meteors an hour; meteor storms can produce that many a second. Meteor showers are well publicized, with tips for best viewing times. No equipment is needed to view a meteor shower, but binoculars help focus on fainter shooting stars. A lounge chair provides comfort for all that looking up. Scan the whole sky; meteors may emerge anywhere. The Great Meteor Storm of November 1833 astonished and awed Western Hemisphere observers with an estimated 240,000 meteors emanating from Leo.

||||||||||||||||||||||||||||||||||||||||||||||||||||||||||||||||||||||||||||||||||||

Asteroids

Rocky bodies left from the formation of the solar system about 4.5 million years ago, asteroids mainly occur in a belt between Mars and Jupiter.

KEY FACTS

A NASA-sponsored program tracks larger asteroids near Earth to gain important information.

+ fact: Asteroids are rich in metals and one day could be mined.

+ fact: Asteroids reveal details of solar system formation and planet migration.

+ fact: Asteroid names include Jabberwock and Dioretsa, a backward-orbiting asteroid.

Hundreds of asteroids are theoretically within range of even a 3-inch telescope. But asteroids are notoriously difficult to spot; their light is faint and they are indistinguishable from stars. Comparing the field of view through a telescope with a sky chart may confirm a starlike object not on the map as an asteroid. Or, you can map what you see through the telescope and compare it with the same view several hours later to determine which objects have moved. Those objects likely are asteroids. The millions of fragments in the asteroid belt collide and sometimes break free from the belt, streaking toward Earth as meteors. In February 2013, an asteroid half the size of a football field came harmlessly within 17,200 miles (27,700 km) of Earth, inside the orbits of some satellites.

Kuiper Belt

A major comet breeding ground, the Kuiper belt is a region of ice and rocky debris that begins just past Neptune and extends outward through the solar system.

KEY FACTS

The Kuiper belt has an outer extension of icy and rocky debris—the scattered disk.

+ fact: The Kuiper belt is about 20 times wider than the asteroid belt between Mars and Jupiter.

+ fact: Kuiper belt objects (KBOs) are difficult to measure because they are so distant.

+ fact: KBOs are also called trans-Neptunian objects, or TNOs.

The Kuiper (kind of rhymes with "wiper") is named for 20th-century Dutch astronomer Gerard Kuiper. It is a fertile field of debris, loaded with objects large and small that divulge information about the history of our solar system. The belt also contains a number of almost planet-size objects. Pluto is the biggest object in the Kuiper belt, at about 1,400 miles (2,300 km) wide.

The demoted planet shares the belt with other dwarf planets Makemake and Haumea (Eris resides in the scattered disk). A number of Kuiper belt dwarf planet candidates are under investigation. Some of these objects seem to be in unusual orbits.

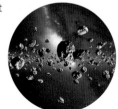

|||||||||||||||||||||||||||||||||||||||||||||||||||||||||||||||||||||||||||||||||||

Oort Cloud

The Oort cloud is an enormous spherical cloud that houses a vast population of icy objects. It encircles the distant outer edge of the solar system.

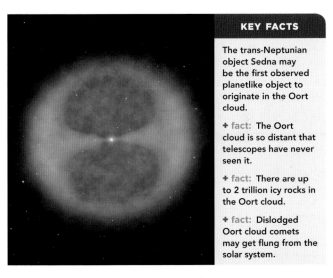

KEY FACTS

The trans-Neptunian object Sedna may be the first observed planetlike object to originate in the Oort cloud.

+ fact: The Oort cloud is so distant that telescopes have never seen it.

+ fact: There are up to 2 trillion icy rocks in the Oort cloud.

+ fact: Dislodged Oort cloud comets may get flung from the solar system.

In the mid-20th century, astronomers Ernst Öpik and Jan Oort theorized the existence beyond Pluto of an even more distant cloud of icy debris left over from the solar system's formation. Now called the Oort cloud, or sometimes the Öpik–Oort cloud, it is one of three solar system nurseries for comets, along with the Kuiper belt and its nearby scattered disk. Comets dislodged from the Oort cloud are long-period comets that can take thousands and even millions of years to orbit the sun. Short-period comets, with orbital periods less than 200 years, originate from the Kuiper belt. The Oort cloud's nearest edge lies some 465 billion miles (750 billion km) away. Its outer edge may extend 20 times farther than that, where the sun has less gravitational influence than nearby stars.

Planets

In 2006, the International Astronomical Union changed the rules for what defines a planet, demoting Pluto and ending more than seven decades of belief in a nine-planet solar system.

KEY FACTS

Current planets are Mercury, Venus, Earth, Mars, Jupiter, Saturn, Uranus, and Neptune; dwarf planets are Pluto, Ceres, Eris, Makemake, and Haumea.

+ fact: In the 1850s, the official list of planets numbered 41.

+ fact: If Earth's moon had an orbit independent of the Earth, it could be a planet.

Discoveries of new objects larger than Pluto, some with their own moons, made it necessary to rethink the definition of a planet. The International Astronomical Union established criteria an object needed to meet to be considered a planet: It must orbit a star, not be a satellite of another object, remain spherical by the pull of its own gravity, and be substantial enough to clear objects from its orbit. Pluto didn't meet the last benchmark. Currently, 13 sky objects qualify as planets or dwarf planets, 8 from the traditional list and 5 dwarf planets, and there is a waiting list of other objects under consideration. The four planets closest to the sun in the inner solar system have solid, rocky natures and are known as terrestrial planets; the four outer ones are known as gas giants.

Planet Viewing

Observing the planets can be done "old school," without gear or charts, or it can involve more effort and a large array of equipment and resources.

KEY FACTS

Some planets can be observed with the naked eye; others require optics.

+ fact: Venus, brightest object in the sky after the sun and moon, is highly visible to the naked eye.

+ fact: Jupiter is also visible; its moons are starlike points through binoculars.

+ fact: Saturn can be viewed unaided; viewing its rings requires a telescope.

Both hardware and software greatly aid planet viewing. Good binoculars and a telescope with a lens aperture of 3 inches (7.5 cm) or more are basic tools to be combined with charts, almanacs, and other resources that help determine the best viewing times. The Internet is a tremendous source of detailed information; Space.com is a great place to start, with its monthly summaries and daily updates of planetary locations. Mobile apps provide another level of assistance, and many are available free or for only a small fee. Add a clear night and a location with unobstructed viewing and minimal light pollution, and you're all set. With the demand for commercial space travel, it is only a matter of time before nonastronauts can add Earth to their planet-viewing wish lists.

Mercury

Smallest and innermost of the planets, Mercury is close enough to be seen with the naked eye but is one of the more difficult planets to spot.

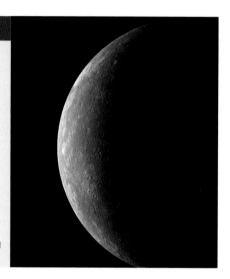

KEY FACTS

Mercury, slightly larger than Earth's moon, has a highly eccentric orbit at irregular distances from the sun.

+ fact: Mercury has a diameter of 3,031 mi (4,878 km).

+ fact: The planet's distance from the sun is 29–43 million mi (47–69 million km).

+ fact: Its orbital period is 88 Earth days, and its rotational period is 58.6 days.

The surface of Mercury is barren, rocky, and pocked with craters. A molten iron core makes up about 65 percent of the planet, which essentially has no atmosphere. Mercury circles the sun every 88 days, making it invisible much of the time due to competition with the sun's glare. The planet is best viewed in the west in evenings in March and April and in the east in mornings in September and October. Extremely clear viewing conditions are usually needed to see it, and it can be a challenge even with binoculars or a telescope. Consult resources such as almanacs and astronomy sites on how best to hunt for it. Mercury is named after the fleet Roman messenger god.

Venus

Neither similarities in size, mass, and orbital year with Earth nor having the goddess of love for a namesake can mask the treacherous nature of Venus.

KEY FACTS

Earth's closest neighbor, Venus has phases somewhat similar to those of Earth's moon, but has no moon of its own.

+ fact: Venus has a diameter of 7,521 mi (12,103 km).

+ fact: The planet has a distance from the sun of 67.2 million mi (108.2 million km).

+ fact: Its orbital period is 225 Earth days, and its rotational period is 243 days.

Although farther from the sun than Mercury, Venus has a greenhouse-type atmosphere that traps the sun's heat and sends temperatures soaring to 860°F (460°C). The atmosphere is thick with carbon dioxide and is blanketed with a 40-mile (64 km) layer of sulfuric acid, which makes Venus highly reflective and bright, but hides its surface. Known as both the "evening star" and the "morning star," Venus is visible for some months each year, low to the horizon in the western sky at nightfall or rising in the east before sunrise. Telescope users generally get a better view at twilight when there is not so much contrast with the dark sky because the planet's brightness can be overpowering.

Transit of Venus

This rare phenomenon, the passing of Venus across the face of the sun, last took place in June 2012. The next transit does not occur until December 2117.

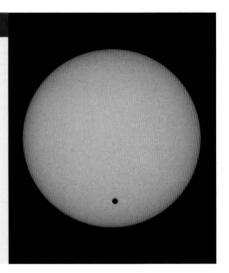

KEY FACTS

The ancient Babylonians may have recorded the transit as early as 3000 B.C.

+ fact: Captain James Cook set up an observation post in Tahiti to take transit measurements in 1769.

+ fact: In 1882–83, John Philip Sousa wrote the "Transit of Venus March."

+ fact: The June 2012 transit was celebrated with observing parties, webcasts, and music.

The alignment that places Venus directly between the Earth and sun affords a chance to see Venus as a dot that slowly moves across the face of the sun. The rare transits occur in pairs eight years apart, with more than a hundred years between pairs. In recent centuries, each transit was observed and recorded with increasingly sophisticated technology. The anticipated transit of 1769 inspired England's Royal Society to send a team, with Captain Cook at the helm of the H.M.S. *Endeavour*, to observe the transit from southern latitudes.

Transits provide an opportunity to learn about the size of the solar system. NASA's webcast of the June 2012 transit of Venus as viewed from Mauna Kea, Hawaii, is archived on its website.

Mars

The stuff of alien fantasies, the Mars of reality is just as fantastic. And we know a lot about it, thanks to concentrated efforts to study the potentially habitable red planet.

KEY FACTS

Mars has two moons called Phobos and Deimos—"fear" and "panic"—and the solar system's largest volcano.

+ fact: Mars has a diameter of 4,222 mi (6,794 km).

+ fact: Mars is 141.6 million mi (227.9 million km) from the sun.

+ fact: Its orbital period is 687 Earth days, and its rotational period is 24.6 Earth hours.

Mars and Earth share a number of similarities. A day on Mars is just a tad longer than an Earth day. Mars has seasons, an atmosphere, clouds, and polar caps, but it is much smaller than Earth. As numerous remote and on-site investigations attest, the surface of Mars is barren and a dusty red, the result of a high iron oxide, silicon, and sulfur content. Its equator is marked by immense volcanoes and dramatic canyons. Mars is visible to the naked eye, but its apparent size and brightness vary significantly during its 687-day orbit. A modest telescope reveals details such as dark lava, boulder fields, and frozen polar caps. Binoculars will show the planet's reddish color.

|||||||||||||||||||||||||||||||||||||||||||||||||||||||||||||||||||||||||||||||

Mars Rovers

Mars has had more than its fair share of scientific inspection in the form of flyby probes, orbiters, landers, and rovers, including the latest—the Curiosity rover.

KEY FACTS

The Mars Pathfinder spacecraft delivered rover Sojourner to Mars in 1997.

+ fact: Rover Opportunity, originally expected to travel 2,000 ft (0.6 km), has racked up more than 22 mi (35 km).

+ fact: Rover Opportunity's twin, Spirit, went off-line in March 2010.

+ fact: Rover Curiosity is about the size of a Volkswagen Beetle.

Robotic vehicles festooned with scientific equipment provide invaluable information about the surface of Mars. The rover Opportunity was deployed in 2004, and was expected by some to survive there only 90 days; it completed its 12th year on Mars in January 2016, providing an ongoing bonus of data. The Curiosity rover, a car-size vehicle carrying a payload of ten of the most sophisticated science instruments ever designed, landed there in August 2012. It is looking at the planet's geology, past planetary processes, surface radiation, and biological potential. In early 2013, it sent back the first nighttime photos of Mars. Yet much of the excitement of Curiosity's mission involves the evidence it has detected of water-bearing minerals in the planet's rocks.

||||||||||||||||||||||||||||||||||||||||||||||||||||||||||||||||||||||||||||||||||||||||||

Asteroid Belt

A gap of 340 million miles (547 million km) between Mars and Jupiter, the asteroid belt marks the division between the solar system's terrestrial planets and its gas giants.

KEY FACTS

The dwarf planet Ceres comprises one-third of the mass in the asteroid belt.

+ fact: Jupiter's orbit contains a group of asteroids known as the Trojan asteroids.

+ fact: Earth has its own small Trojan asteroid belt located at about 60° east or west of the sun.

+ fact: Each of the Beatles has an asteroid named for him.

Jupiter's massive gravitational field prevents the rocky debris in the asteroid belt—leftovers from the formation of the solar system—from coalescing into planets. The asteroids in the belt are like a warehouse of raw materials that under different circumstances could have made the transition to full-fledged planets. The debris won't achieve that destiny, but larger objects in the belt, such as Ceres, have the chance to be considered dwarf planets. Astronomers have cataloged more than 120,000 bodies in the asteroid belt, and more than 13,000 are named, sometimes very personally for girlfriends and favorite writers. According to the guidelines set out by the International Astronomical Union, near-Earth objects are supposed to have names taken from the mythology of any world culture.

Jupiter

A planet of superlatives, Jupiter is the largest by far and has the largest number of satellites, with 67 moons, making it a kind of solar system unto itself.

KEY FACTS

Jupiter has a ring, but it is small and thin compared to other planets' rings.

+ fact: Jupiter has a diameter of 88,846 mi (142,984 km).

+ fact: The planet's distance from the sun is 483.7 million mi (778.4 million km).

+ fact: Its orbital period is 11.9 Earth years, and its rotational period is 9.9 Earth hours.

Jupiter has a massive influence in the solar system with its strong magnetic field and gravitational pull, intense emissions of radio waves, and bursts of radiation. It has a metallic core but otherwise is a swirling mass of gases: hydrogen, helium, methane, and ammonia. Its complete rotation lasts just short of ten hours; at this clip, the atmosphere appears to us as faint pastel bands. Persistent storms, including the Great Red Spot, a perpetual high-pressure zone, add to the spectacle.

At the other extreme, Jupiter takes nearly 12 years to orbit the sun, spending nearly a year in each constellation of the zodiac and becoming an easy target for sky-watchers.

Jupiter's Moons

In 1610, Galileo spotted the four largest moons of Jupiter with his telescope. Io, Callisto, Ganymede, and Europa became known as the Galilean moons.

KEY FACTS

Jupiter has a chart-topping 67 moons; currently, some of them still lack names.

+ fact: Callisto is the most heavily cratered object in the solar system and lacks a molten core.

+ fact: Moon Io is the most volcanically active object in the solar system.

+ fact: Many of Io's more than 400 volcanoes erupt almost continuously.

Jupiter's intense gravitational pull, second only to the sun's, has lured many objects into its orbit. It governs these moons like the masterful planet that it is, and its four largest moons do the planet proud. Jupiter's Galilean moons are large enough that if they orbited the sun, they would qualify as planets. Among them, Ganymede is the largest moon in the solar system, larger than Mercury, Ceres, Pluto, and Eris combined. It generates its own magnetic field.

Europa may harbor a saltwater ocean under its icy surface, and it may have twice as much water as Earth does. The Galilean moons can be viewed easily from Earth with binoculars. In orbit, all the moons keep the same face toward Jupiter, just as Earth's moon does.

Saturn

Gas giant Saturn, the second largest planet, is best known for its rings, but it has also acquired a multitude of moons. Despite its massive dimensions, Saturn is less dense than water.

KEY FACTS

Saturn with its rings is the most recognizable planet.

+ fact: Saturn has a diameter of 74,898 mi (120,536 km).

+ fact: Its distance from the sun is 885.9 million mi (1.4 billion km).

+ fact: Saturn's orbital period is 29.4 Earth years, and its rotational period is 10.7 Earth hours.

Known to the ancient Mesopotamians, Saturn was dubbed "the old sheep" for taking more than 29 years to orbit the sun. The plodding orbit is countered by speedy rotation; a day on Saturn lasts just under 11 hours. And although the planet's volume could contain 763 Earths, its low density means that it would float in water if a big enough basin could be found. It also means that the planet appears considerably squashed in profile; the diameter through the poles is significantly shorter than its equatorial diameter. Saturn is visible to the naked eye, but binoculars or a telescope are needed to view details. The Cassini orbiter has captured some amazing images.

Saturn's Rings

Saturn's iconic rings are composed of rubble, dust, and ice. The rings may also contain the pulverized remains of moons, comets, and asteroids that strayed too close.

KEY FACTS

Galileo viewed Saturn through his telescope in 1610, but he thought the side bulges he saw were separate bodies.

+ fact: Saturn's rings vary in width and thickness; they extend 170,000 mi (274,000 km) from side to side.

+ fact: Saturn's rings were named alphabetically in the order of their discovery.

+ fact: Saturn's rings may disappear.

A ring around Saturn was first proposed by Dutch astronomer Christiaan Huygens in 1655. We now know that Saturn has multiple rings fashioned from billions of icy particles—some as small as grains of sand and others as big as buildings. The rings are complex and not solid, containing ringlets and gaps between them. The appearance of the rings, visible with a small telescope, changes over time. When the tilt of Saturn is toward Earth, the rings are visible from a broad top-down view.

Roughly every 14 years, the rings are tilted edgewise toward Earth, making them all but disappear from view—an event that will next occur in 2023. The Cassini-Huygens probe that traveled to Saturn's largest moon was named for an Italian-French and a Dutch astronomer.

Saturn's Moons

Saturn's impressive assemblage of moons includes many that exhibit unusual features. Only 53 of the 62 known moons currently have names.

KEY FACTS

Saturn's largest moon is the massive Titan, larger than the planet Mercury.

+ fact: Titan's atmosphere contains more than 90 percent nitrogen.

+ fact: Moons Titan and Enceladus have the potential for life.

+ fact: Some of Saturn's moons share orbits, traveling the same paths.

Among Saturn's fascinating moons, Titan has surface liquids in the form of two large methane lakes. The poles of Enceladus shoot geysers of water vapor and ice particles that suggest liquid water beneath its frigid surface. Moon Mimas sports a large impact crater of unknown origin. Several moons appear to wrangle the billions of ice particles that comprise Saturn's rings. These "shepherd moons"— Pan, Atlas, Pandora, and Prometheus—straddle two of the rings and keep them intact. Much of what we know about Saturn, its moons, and its rings comes from the data being collected by the Cassini orbiter since its arrival at Saturn in 2004. Cassini transmits many raw images that are made available to the public before they are calibrated and analyzed by NASA scientists.

Uranus

Uranus is a tilted planet, spinning on its side at an angle of 98° to its orbital plane. It was probably knocked off-kilter by a massive collision.

KEY FACTS

Uranus has a retro-grade rotation, revolving in the opposite direction of most other planets.

+ fact: Uranus has a diameter of 31,764 mi (51,119 km).

+ fact: The planet's distance from the sun is 1.8 billion mi (2.9 billion km).

+ fact: Its orbital period is 84.02 Earth years, and its rotational period is 17.24 Earth hours.

Distant Uranus, barely visible to the naked eye, was incorrectly identified as a star by ancient astronomers before its discovery in 1781 by astronomer William Herschel. Spotted more realistically through binoculars, Uranus appears as a small blue-green disk, the color due to methane gas in its upper atmosphere. Uranus is considered an ice giant, with temperatures reaching down to −357°F (−216°C). Uranus's tilted rotation means that the planet's poles point toward the sun: first one pole and then, 42 years later, the other. In addition to 5 large moons and more than 20 smaller ones, 13 narrow rings surround Uranus. The ring system was imaged by the Hubble Space Telescope.

Neptune

Named for the Roman god of the sea, Neptune displays a vivid blue color, a product of the planet's surrounding clouds of icy methane.

KEY FACTS

Neptune's moon, Triton, has a retrograde orbit, orbiting opposite the direction of Neptune's rotation.

+ fact: Neptune has a diameter of 30,599 mi (49,244 km).

+ fact: The planet's distance from the sun is 2.8 billion mi (4.5 billion km).

+ fact: Its orbital period is 164.79 Earth years, and its rotational period is 16.11 Earth hours.

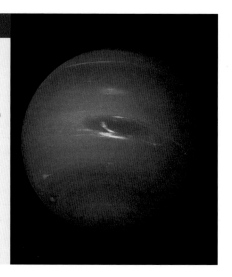

The outermost major planet in the solar system, Neptune is an ice giant similar in size to Uranus, although denser. A deep atmosphere containing hydrogen, helium, and a small amount of methane surrounds it. The planet seems to have an internal heat source and a puzzling tendency toward very turbulent weather, despite its great distance from the sun. Winds on Neptune can reach more than 1,300 miles (2,100 km) an hour.

Neptune is surrounded by a system of rings that are faint and lumpy and appear to be younger than the planet itself. The rings could be the remnants of a former moon pulled apart by Neptune's gravity.

Dwarf Planets & Plutoids

Dwarf planets are almost planets, which tend to be small and roundish and travel with debris in their orbits. Plutoids are dwarf planets in the trans-Neptunian Kuiper belt.

KEY FACTS

Haumea is named for a mythical Hawaiian sorceress; its moons are named for her daughters.

+ fact: Haumea is sometimes called the "cosmic football" for its shape and motion.

+ fact: Makemake is named for the chief god of Easter Island in the South Pacific.

+ fact: Eris was originally called Xena after the TV warrior princess.

When Pluto was demoted in 2006, it was put into a category of dwarf planet. Besides Pluto, several other objects in the Kuiper belt qualify. Generally, dwarf planets are larger than Mercury, roundish, and are not satellites of other objects. Haumea, a rocky, egg-shaped object, tumbles end over end as it spins every four hours. Makemake seems to have an atmosphere containing nitrogen and methane. It shines red and has no moon. The orbit of Eris and its moon, Dysnomia (Greek for "lawlessness"), takes it 10 billion miles (16 billion km) beyond the Kuiper belt. Dwarf planet Ceres is not a plutoid; it appears in the interior asteroid belt beyond Mars. Astronomers are investigating hundreds of objects to determine whether the objects qualify for dwarf planet status.

Pluto

Pluto enjoyed 76 years of planetary status until, one day in 2006, it didn't. Demoted to dwarf planet, the celestial body named for the god of the underworld retains legions of fans.

KEY FACTS

Some astronomers consider Pluto and its moon Charon to be a double planet.

+ fact: Pluto has a diameter of 1,473 mi (2,370 km).

+ fact: The planet's distance from the sun is 3.7 billion mi (5.9 billion km).

+ fact: Its orbital period is 247.92 Earth years, and its rotational period is 6.39 Earth days.

In 1930, an assistant at the Lowell Observatory found a long-anticipated planet beyond Neptune. Excitement about the newbie generated a naming contest, which an 11-year-old British girl won. She proposed the name of the Greek god of the underworld that coincidentally was the name of Mickey Mouse's popular pup. By 2006, the International Astronomical Union, or IAU, had its doubts about Pluto and demoted it to the status of dwarf planet. Pluto did not meet one of the prerequisites the IAU had drawn up: It didn't clear its orbit of debris. Pluto now lends its name to dwarf planets that inhabit the Kuiper belt—the plutoids. Many of Pluto's devoted fans mourn its demotion.

Ceres

Ceres, a dwarf planet, rose out of the asteroid belt between Mars and Jupiter. Since its discovery in 1801, Ceres has worn several solar system hats: planet, asteroid, and now dwarf planet.

KEY FACTS

Ceres has a rocky inner core and a mantle of water ice, and is one of two dwarf planets without a moon.

+ fact: Ceres's diameter is 590 mi (950 km).

+ fact: The dwarf planet's distance from the sun is 258 million mi (415 million km).

+ fact: Its orbital period is 4.6 Earth years, and its rotational period is 9.08 Earth hours.

In 2006, Ceres was promoted to the new group of dwarf planets. About one-fourth the diameter of Earth's moon, Ceres is the largest object in the asteroid belt, an area holding thousands of objects that are remnants left from the formation of the solar system. Dwarf Ceres is the fifth planet from the sun and is classified among the terrestrial planets because of its rocky nature. It is suspected of having water ice under its crust because it is less dense than Earth and its surface shows evidence of water-bearing minerals. Named for the Roman goddess of agriculture, Ceres takes 4.6 Earth years to complete one orbit around the sun. Ceres was discovered by Italian astronomer Giuseppe Piazzi.

The Universe

The word "universe" evokes all the superlatives one can imagine, only to take on more with each breakthrough in cosmic understanding enabled by sophisticated technology.

KEY FACTS

Many physicists and astronomers believe that our universe is just one of an unknown number of universes.

+ fact: The age of our universe is about 13.7 billion years old.

+ fact: The first stars were formed about 100 million years after the big bang.

+ fact: The first galaxies formed less than 600 million years after the big bang.

The unfolding story of the universe involves greater stretches of the imagination than most science fiction tales could devise. Consider the plot: An infinitely small point of infinite density and infinite gravity erupts to create both space and time, unleashing radiation that eventually forms matter that eventually forms stars, galaxies—and us. The initial radiation of the universe's chaotic beginnings was left by hot plasma (ionized gas) that emitted a microwave signal. We created a complete map of the microwave sky, the oldest light in the universe, using space probes such as the Wilkinson Microwave Anisotropy Probe. The probe accomplished a full-sky survey of background microwave radiation between 2001 and 2010, which scientists scrutinize for information about this phenomenon.

||||||||||||||||||||||||||||||||||||||||||||||||||||||||||||||||||||||||||||||||||||||||

Big Bang

More accurately described as an expansion than an explosion, the big bang set in motion the formation of the universe about 13.7 billion years ago.

KEY FACTS

Events immediately after the big bang occurred at an incomprehensibly fast speed.

+ fact: In 10^{-35} seconds after the big bang, its energy turned into matter.

+ fact: In 10^{-5} seconds after the big bang, the universe's natural forces took shape.

+ fact: In 3 seconds after the big bang, the nuclei of simple elements formed.

The term "big bang" was coined in the 1950s by British astronomer Fred Hoyle. Ironically, he used it derisively, but it stuck. Current thinking holds that the universe began as a singularity, a point of infinite gravity, where space, time, and all subsequent matter and energy were contracted into an object without size. At the point in which contents of the singularity escaped, a dramatic expansion began that has played out over billions of years. Radiation from the initial event persists as detectable cosmic background radiation. The origin of the singularity remains scientifically elusive, but predicted scenarios of the universe's demise include a big crunch, a big chill, or a big rip. Astronomers have a few tens of billions of years to look for answers.

Galaxy

Gravity holds together the vast assemblage of stars, interstellar dust and gas, and dark matter (little understood matter that emits no light) that form a galaxy.

KEY FACTS

There may be one hundred billion galaxies in the universe.

+ fact: Galaxies occur in clusters throughout the universe.

+ fact: Galaxy clusters form larger associations, giving the universe a "clumpy" nature.

+ fact: The most distant galaxy yet identified is more than 13 billion light-years away.

Shape determines galaxy classification. Spiral galaxies, by far the most numerous, rotate around a bright nucleus. Arms spiral out from that point. Elliptical galaxies may be nearly spherical or stretched to an oblong; stars in them are mostly older red giant stars. Irregular galaxies, which lack a coherent shape, are amorphous collections of stars that include star-forming nebulae. The most recently identified type is the starburst galaxy, so called because the stars in it seem to burst out. Galaxies are so massive that the strength of one can rip apart another, and they can collide even in the relative emptiness of space. The Hubble Space Telescope's website *(www.hubblesite.org)* features an extensive gallery of amazing, downloadable photos that demonstrate galaxy diversity—and beauty.

Milky Way

Home sweet home, the Milky Way galaxy contains our solar system, just a tiny enclave in the vast spiral galaxy of several hundred billion stars.

KEY FACTS

The Milky Way is composed of some 200 to 500 billion stars.

+ fact: Our solar system is located about 25,000 light-years from the galaxy core.

+ fact: The Milky Way's core contains a supermassive black hole.

+ fact: In ancient times, the Milky Way was a more prominent feature in the sky and many myths developed around it.

Thought to be about 13 billion years old, the Milky Way is a bit younger than its oldest stars, which are estimated to be about 13.5 billion years old. The galaxy's disk shape bulges at the center with orange and yellow stars. A corona of old stars and globular clusters extends above and below the disk. The younger, brighter stars and nebulae crowd the arms of the galaxy. The entire spiral structure spins, completing a rotation approximately every 200 million years. Overall, the Milky Way has a diameter of about 100,000 light-years; its thickness ranges from 1,000 to 13,000 light-years. Dark nights and clear skies give the best chance of viewing the Milky Way. In the Northern Hemisphere, the summer months usually offer the best views of the galaxy.

Andromeda Galaxy

A faint oval smudge in the gap between the constellations Cassiopeia and Andromeda, the Andromeda galaxy is the most distant object visible to the naked eye.

KEY FACTS

Astronomer Edwin Hubble's study of the Andromeda galaxy was the first confirmation of the existence of galaxies other than the Milky Way.

+ fact: The light we see from the Andromeda galaxy left the galaxy 2.5 to 3 million years ago.

+ fact: The galaxy has a magnitude of 4.4, making it one of the brighter Messier objects.

Possessing a spiral shape like the Milky Way, the Andromeda galaxy is the larger of the two, containing an estimated one trillion stars. Nevertheless, recent observations indicate that the Milky Way may contain greater mass, if dark matter is included. The Andromeda galaxy is accompanied by several smaller galaxies, including M32 and M110, visible through binoculars. Under good conditions, especially during autumn, the Andromeda galaxy is visible to the naked eye. Astronomers as far back as at least A.D. 964 observed and recorded the feature. Many observers comment that its beauty compares to its namesake, the mythical princess Andromeda. A distance "only" about 2.5 million light-years away turns the Andromeda galaxy into the girl next door.

Galaxy Collisions

Despite the general emptiness of space, galaxies have a tendency to move toward each other and collide, a mechanism that allows galaxies to increase in size.

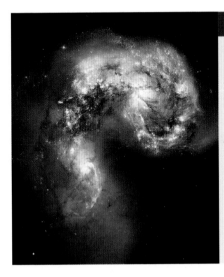

KEY FACTS

A merger from a galaxy collision can take millions of years.

+ fact: When a large galaxy merges with a smaller one, the larger one usually keeps its shape.

+ fact: Galaxy collisions cause friction that can trigger shock waves that lead to star formation.

+ fact: Galaxy collisions provide vital information about galaxy evolution.

In the scheme of the universe, galaxy collisions occur with different frequencies. More often, larger galaxies collide with dwarf galaxies and incorporate the mass of the dwarf galaxies into their own. Collisions of larger galaxies happen less frequently. Collisions can produce a galaxy of a different shape; two spiral galaxies, for example, can form an elliptical galaxy.

The Hubble Space Telescope has revealed details of the approaching collision—four billion years from now—between the Milky Way and the Andromeda galaxy. The news seems good for our solar system: displacement, perhaps, but not destruction. This future merger of the two galaxies sometimes is referred to as Milkomeda or even the Andromeda Way.

Nova

A spectacular sudden surge in the brightness of a star, a nova is the equivalent of a supermassive atomic explosion, but one from which the star often can recover.

KEY FACTS

Novae are not often spotted, but when they are, you can find them with binoculars along the Milky Way.

+ fact: At times, a nova can be viewed with the naked eye.

+ fact: The density of a white dwarf allows it to draw in hydrogen from the larger star.

+ fact: A classical nova erupts only once.

Typically, a nova occurs when stars in a binary system reach different points in their stellar evolution, with one star collapsed into a dense white dwarf and the other in a red giant phase. The white dwarf siphons off hydrogen from the red giant, and the temperature and pressure rise until the gas erupts in a thermonuclear explosion. This causes the white dwarf to brighten for a period of hours or days by as many as 10 magnitudes before dimming over a period of months as the remnants of the blast dissipate. Novae are not necessarily fatal to the stars involved and often occur in cycles in some binary pairs. The Keck Interferometer, a two-telescope system at the W. M. Keck Observatory in Mauna Kea, Hawaii, has captured invaluable data about a nova in the constellation Ophiuchus.

III

Supernova

Unlike the nova, a supernova is a one-time event. It signals the death of a star, an irrevocable dissolution and transformation of its material.

KEY FACTS

A supernova can briefly release as much energy as all the stars in the Milky Way.

+ fact: A supernova creates raw material for the formation of new stars.

+ fact: The Crab Nebula (M1) in the constellation Taurus is the remnant of an event in 1054.

+ fact: The brightest supernova in recent history occurred in 1987.

Like the nova, a supernova can also occur in a white dwarf if its core takes in so much hydrogen that it implodes. A supernova also announces the end stage for a red supergiant when its core implodes from intense heat and pressure, unleashing a violent shock wave that blasts away the surrounding cloud of gas. The supernova was first recorded by the Chinese in A.D. 185. Supernovae again were recorded in 1006, 1054, and 1572, and in the modern era. The bright flash of a supernova can occasionally be viewed by the naked eye, but usually requires an 8-inch or larger telescope. More than 25 years after its explosion in the Large Magellanic Cloud, Supernova 1987A continues to glow as it transforms into a supernova remnant, which the Hubble Space Telescope has imaged.

Black Hole

A black hole provides a one-way ticket to visual oblivion for the contents of a supernova explosion. The resulting gravitational field collapses in on itself, allowing no light to escape.

KEY FACTS

The point of no return at the edge of a black hole is called the event horizon.

+ fact: The energy from stellar gases swirl into a black hole like a whirlpool.

+ fact: The black hole in a galaxy in the con-stellation Virgo seems to have the density of 3 billion suns.

+ fact: The Chandra X-ray Observatory pro-vides critical data on black holes.

Born in one of the universe's most dramatic and awe-inspiring events, a black hole is one of the two options that follow a supernova explosion. The last explosive gasps of a red supergiant create a gravitational field so intense that the force of gravity causes matter to collapse inward. This forms the black hole, in which distance between sub-atomic particles reduces to zero, and density and gravity expand to their maximum expression. Black holes affect the stars and gas around them, which is one way they can be detected. Small black holes formed at the creation of the universe, and supermassive black holes formed when the galaxies they inhabit were formed. Supermassive black holes spin furiously as they swallow stars and merge with other black holes.

Quasar

The bright light of a quasar is fed by a black hole in the form of a brilliant stream of energy that a black hole releases as it consumes the contents of nearby stars.

KEY FACTS

Coined in the 1960s, the word "quasar" comes from "quasi-stellar," as in quasi-stellar radio source.

+ fact: More than 120,000 quasars have been identified.

+ fact: Quasars are among the universe's very ancient objects.

+ fact: The Sloan Digital Sky Survey catalogs positions and absolute brightness of celestial objects such as quasars.

Bright, deep sky objects with a starlike appearance, quasars puzzled astronomers from the time they were first detected from their radio emissions. When spotted, quasars seemed too bright to be so far away. A boost from association with a black hole was proposed for the anomaly. When quasars were found to be abundant, they came to be considered a part of normal galactic evolution. Though distant, a quasar is within the reach of a skilled amateur astronomer. The quasar 3C 273 can be spotted some 3 billion light-years away through an 8-inch-aperture telescope just 5 degrees northwest of the figure's head in Virgo. Sometimes, though, even the Hubble Space Telescope can struggle to observe a quasar embedded in a galaxy obscured by a large amount of cosmic dust.

Binary & Multiple Stars

"Twinkle, twinkle, little stars"—plural—might be more appropriate, because about 80 percent of stars exist with a companion or companions.

KEY FACTS

The star pair Mizar and Alcor in Ursa Major are about 3 light-years apart and are not gravitationally linked.

+ fact: Mizar is actually a double binary, and Alcor is a binary, making a 6-star system.

+ fact: Binary stars, each star with its own orbiting planet, have been discovered.

+ fact: Binary stars, each with multiple planets in orbit, have also been discovered.

Our singleton sun aside, the massive gas clouds that nurture stars usually produce them in pairs or greater multiples that remain gravitationally attached to each other throughout their lives. Binary stars, triples, and larger groups are the norm, whether the companion is a twin of similar size and luminosity or a group of small siblings that stick close to a more dominant star. Double stars can be optical doubles, which only appear to be close, or true binary stars, with two stars orbiting a common center of gravity. Powerful binoculars or a telescope are often needed to pick out the individual members of a binary pair. Alpha Centauri is a binary star that seems to be bound to Proxima Centauri, which is far from Alpha Centauri but closer to us, as a so-called wide binary.

Nebula

Giant clouds of stellar gas, nebulae provide the raw materials for stars, planets, and galaxies. They are present at both star birth and star death.

KEY FACTS

Despite the name, a planetary nebula does not have an association with planets.

+ fact: Dark nebulae are so dense that they hide the light of stars within them, not only behind them.

+ fact: The Helix Nebula is the closet nebula to Earth.

+ fact: The North American Nebula is a large emission nebula in the constellation Cygnus.

Various types of nebulae represent those giant gas clouds in their different and dynamic roles and relationships with stars. Emission nebulae are star-forming clouds energized by the young stars within them, such as the middle star of Orion's belt. Planetary nebulae are formed of the gas departing from a dying red giant, as seen in the Ring Nebula in the constellation Lyra. Dark nebulae are collections of interstellar dust that obscure the light of the stars behind them. Reflection nebulae shine from the light of nearby stars. Many nebulae are visible with a 4- or 6-inch telescope, maximized with higher magnification and filters. The Orion Nebula, which is 1,500 light-years from Earth, is considered a textbook example of a planetary nebula, showing a number of different star-forming processes.

Eagle Nebula

The Eagle Nebula, located in the tail portion of the constellation Serpens about 7,000 light-years away, is an active region of star formation.

KEY FACTS

The dark towers are 56 trillion mi (90 trillion km) high.

+ fact: Images from the Chandra X-ray Observatory show that the pillars are low in x-ray content.

+ fact: The scant content perhaps signals the pillars' end.

+ fact: Further study suggests that a supernova destroyed the pillars 6,000 years ago, but evidence hasn't yet reached Earth.

The stellar nursery in the Milky Way known as the Eagle Nebula is actually a combination nebula and star cluster. Philippe Loys de Chéseaux discovered it between 1745 and 1746, as did Charles Messier independently in 1764. The nebula has dark pillars of dense material that rise at its center and can be seen with a 12-inch telescope. The Hubble Space Telescope captured these pillars in a now iconic image known as the Pillars of Creation. Newborn stars sculpt the pillars by burning away some of the gas within the nebula. The nebula in general can be viewed with low-powered telescopes and even binoculars. But to view the Eagle Nebula in all its dramatic glory, you will want to check out the online images available on HubbleSite (www.hubblesite.org).

||||||||||||||||||||||||||||||||||||||||||||||||||||||||||||||||||||||||||||||||||

Exoplanets

Exoplanets, short for extrasolar planets, are celestial bodies beyond our solar system that orbit stars and share other characteristics of known planets.

KEY FACTS

More than 2,300 exoplanets have been identified, with many more expected.

+ fact: Exoplanets include "hot Jupiters," gas giants that orbit closer to their stars than Mercury does to the sun.

+ fact: Hot Jupiters are also known as "roaster planets."

+ fact: Super-Earths refer to rocky exoplanets about twice the mass of Earth.

Radio astronomers discovered the first extrasolar planet in 1991. Since then, ground-based and space-based telescopes have confirmed the presence of hundreds of objects that orbit stars of other systems or that freelance as nomads—planets "kicked out" of their star systems. Various scientific organizations use different criteria to confirm an exoplanet's status, coming up with different totals, but all concede that the numbers are increasing. The ultimate find among exoplanets would be an Earthlike body in the habitable zone of its solar system that had not only a potential for life but also signs of life itself. To keep up with the latest exoplanet news, visit *exoplanets.nasa.gov*. It documents exoplanet research at the Jet Propulsion Laboratory.

II

Kepler Space Telescope

The Kepler Space Telescope launched in 2009 on a mission to monitor more than 150,000 stars, looking for Earthlike exoplanets. It has confirmed more than 2,000.

KEY FACTS

The Kepler instrument is a photometer (light meter) telescope with a large field of view.

+ fact: The telescope observes a large field of stars over time, continuously monitoring the brightness of all.

+ fact: Kepler has found a 6-planet system orbiting an 8-billion-year-old star at 2,000 light-years away.

+ fact: In 2014 Kepler began its next mission, K2.

The Kepler Space Telescope scrutinizes our portion of the Milky Way galaxy for extrasolar planets and the stars they orbit. The big prize would be an Earthlike planet technically capable of supporting life (known as a "just right Goldilocks" planet), but all kinds of exoplanets are given attention: gas giants, hot super-Earths with short-period orbits, and ice giants. Kepler finds potential candidates through the concept of the transit—a planet crossing in front of its star from the observer's point of view. When this happens, the star's brightness dims by a factor related to the size of the object. A periodic occurrence at four equal intervals confirms an object as a planet candidate. Follow-up observations are made to eliminate other possibilities before the planet is verified.

Dark Energy

Einstein rightly realized that space was not just empty space. Though only inferred, dark energy may be the "dark force" involved in rapid expansion of the universe.

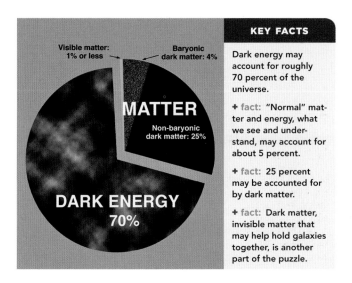

KEY FACTS

Dark energy may account for roughly 70 percent of the universe.

+ fact: "Normal" matter and energy, what we see and understand, may account for about 5 percent.

+ fact: 25 percent may be accounted for by dark matter.

+ fact: Dark matter, invisible matter that may help hold galaxies together, is another part of the puzzle.

Too much was going on in the universe for there not to be something in space besides space. When a study of supernovae showed an unexpected acceleration of the expansion of the universe, some force had to be held accountable. That force has been designated as dark energy, a concept that creates more questions than it answers at present. It can be seen as the reverse of gravity: Dark energy pushes objects away from each other rather than pulling them together and coalescing them. How it operates keeps astrophysicists in a quandary, because it calls into question the laws of gravity and the theory of relativity. Observations by the Hubble Space Telescope in 1998 set this discussion in motion, but a lot more data obviously are needed to arrive at some answers.

Constellations

Over the millennia, stargazers have teased out shapes of humans, animals, and objects from patterns of stars in the night sky. A recognized star pattern is called a constellation.

KEY FACTS

Ptolemy's list of constellations appeared in his astronomical treatise known as the *Almagest*.

+ fact: The name "Almagest" comes from a later Arabic title for the book meaning "the greatest."

+ fact: The constellation Orion, visible on the celestial equator, has inspired many elaborate mythologies.

The International Astronomical Union (IAU) recognizes 88 official constellations. These include the original 48 constellations listed by Ptolemy in the second century A.D., which have names and backstories rooted in Greco-Roman mythology. The Greek constellation Argo Navis has since been divided into three separate constellations. The list also includes a series of "modern" constellations that have been added since about 1600 to fill in gaps, especially in the southern sky. According to the IAU, a constellation refers not just to the star pattern itself, but also to an agreed-upon, bounded segment of the sky in its vicinity. The convention helps us locate objects in the night sky and allows astronomers to convey information about deep-space objects that are not part of a constellation.

Asterism

When is a constellation not a constellation? When it's an asterism, a small group of stars that makes a recognizable, attention-grabbing shape in the sky.

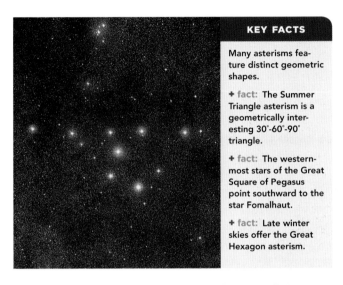

KEY FACTS

Many asterisms feature distinct geometric shapes.

+ fact: The Summer Triangle asterism is a geometrically interesting 30°-60°-90° triangle.

+ fact: The westernmost stars of the Great Square of Pegasus point southward to the star Fomalhaut.

+ fact: Late winter skies offer the Great Hexagon asterism.

A look at the official list of 88 constellations will show that some prominent celestial names, like the Big and Little Dippers, are missing. These well-known star formations are asterisms, small groups of stars that have a distinct shape and may form part of a constellation but are not constellations themselves. The iconic Big and Little Dippers, for example, are contained within Ursa Major and Ursa Minor, respectively. Asterisms are useful points of orientation in the sky and may include stars from multiple constellations. The Winter Triangle, for example, connects the alpha stars Betelgeuse of Orion with Procyon in Canis Minor and Sirius in Canis Major. Similarly, the Summer Triangle asterism connects the bright stars Vega in constellation Lyra, Deneb in Cygnus, and Altair in Aquila.

Cluster

Two kinds of star clusters appear in the deep sky. These dramatic formations are sometimes visible to the naked eye, but a telescope reveals their scale and complexity.

KEY FACTS

In an open cluster, the individual stars often can be resolved through a telescope.

+ fact: Individual stars often escape open clusters and travel on their own.

+ fact: Globular clusters are prominent in the central bulge of the Milky Way.

+ fact: Many of the globular clusters in our galaxy have highly eccentric orbits.

Star clusters are classified as either open or globular clusters. Open clusters are loose concentrations of a few to several thousand associated stars that are bound weakly by gravity. They also share a common star birth "cloud," or source of raw materials. A good example is the Pleiades in the constellation Taurus, a young open cluster that will eventually disperse over time. Another type is the globular cluster, a massive star association that may contain up to a million stars and may be up to 13 billion years old. About 150 of them have been identified, mostly in the outer reaches of the Milky Way. Star clusters in proximity to each other appear to merge, creating larger clusters. Astronomers hope that the James Webb Space Telescope will reveal details of such interactions.

The Pleiades

Any Subaru owner should recognize this asterism, which appears in stylized fashion as the logo for the Japanese auto manufacturer named for the star cluster.

KEY FACTS

The Pleiades cluster (M45) contains hundreds of stars; the 6 or 7 brightest stars make up the famous Seven Sisters asterism, most of them visible to the naked eye.

+ fact: Months best viewed: January–February

+ fact: Seasonal chart location: winter, southwest quadrant

+ fact: Noted skymark: Taurus

Also known as the Seven Sisters, the Pleiades cluster is one of the most easily identified objects in the night sky. A line traced from alpha stars Betelgeuse in Orion through Aldebaran in Taurus brings the asterism into view. The Pleiades' seven bright stars feature prominently in many world mythologies; they get their names from seven sisters in Greek mythology, daughters of the Titan Atlas and the Oceanid Pleione. Zeus set them in the sky to thwart Orion's advances. The brightest star is Alcyone, with a magnitude of 2.86. A Kiowa Indian tale tells of seven sisters chased by giant bears onto a tall rock and then whisked into the sky. The rock is Devil's Tower, a volcanic formation in northeastern Wyoming, and its long cracks are the claw marks of the frustrated bears.

Big Dipper

The first asterism many people learn to locate, the Big Dipper is an abiding feature of the northern sky in all seasons, because of its proximity to the celestial north pole.

KEY FACTS

This 7-star asterism, which forms a distinct ladle or dipper, is incorporated into the constellation Ursa Major. Its orientation changes according to time of the year.

+ fact: Months best viewed: March–April

+ fact: Seasonal chart location: spring, center of chart

+ fact: Noted sky-mark: Ursa Major

To locate the Big Dipper, face the north horizon. As it circles the celestial pole during the year, the asterism faces downward in spring and upward in autumn. The Big Dipper also points to its counterpart in Ursa Minor, the Little Dipper. Follow an imaginary line from the front edge of the Big Dipper's ladle to find Polaris on the end of the Little Dipper's handle. Visible year-round, the Big Dipper served as a constant guide for escaped slaves traveling along the Underground Railroad and is mentioned in many stories and songs, including "Follow the Drinking Gourd." It is known as the Plow in Britain and the Great Cart in Germany. In India it is the Seven Sages. Cultures in much of the world have given locally significant names to this familiar asterism.

Polaris

Not the brightest star in the sky—as many assume—Polaris is actually the 50th brightest star, but its year-round presence makes it an invaluable "skymark."

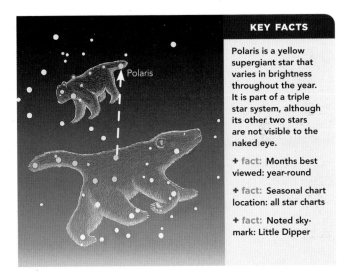

Located directly above Earth's northern axis, Polaris appears to be the star around which the rest of the sky revolves. The stars in the front edge of the Big Dipper asterism line up with it, making the North Star, or polestar, easy to find. Due to the wobbling of the Earth on its axis as an effect of gravitational pull from the sun, moon, and planets—a motion known as precession—the Earth's pole points to different stars over time. The pole will be closest to Polaris about the year 2100, and after that will move through a series of new polestars and reach Vega in Lyra about 12,000 years from now. In the southern sky, Beta Hydri in the southern constellation Hydrus (different from the northern sky constellation Hydra) is the closest bright star to the celestial south pole.

||||||||||||||||||||||||||||||||||||||||||||||||||||||||||||||||||||||||||||||||

Northern Sky

As most of the world's population lives in the Northern Hemisphere, the northern sky has gotten the lion's share of attention, from the ancient astronomers onward.

KEY FACTS

Northern sky constellations are named mostly for figures in Greco-Roman mythology.

+ fact: Some constellations are visible in the Northern Hemisphere all year.

+ fact: Ursa Major is an example of a north circumpolar constellation.

+ fact: Others near the celestial equator are visible in northern and southern skies.

When Greek astronomer Ptolemy codified the constellations in the second century A.D., he followed traditions handed down from ancient Mesopotamian observers who had watched the skies at about latitude 35° N. They would have seen the northern sky up to the celestial north pole and, in theory, a good bit of the southern sky, although objects near the horizon are difficult to see. Constellations and features farther south would have been blocked by Earth. This partially accounts for the fact that ancient astronomers limited their constellation list to 48, although they saw some of the southern constellations some of the time. The "Tonight's Sky" feature at HubbleSite *(www.hubblesite.org)* offers a preview of constellations, deep sky objects, planets, and events visible each month in the skies of the Northern Hemisphere.

||||||||||||||||||||||||||||||||||||||||||||||||||||||||||||||||||||||||||||||||||

Southern Sky

Not fully appreciated until voyages of exploration sailed southern routes beginning in the 1500s, the southern sky harbors extraordinary sights.

KEY FACTS

Crux, or the Southern Cross, is the smallest in area of all the constellations.

+ fact: The Southern Cross appears on the flags of Australia and New Zealand.

+ fact: Many beautiful globular sky clusters are visible in the southern sky.

+ fact: The brightest part of the Milky Way passes over southern skies in winter.

The skies of the Southern Hemisphere contain a glorious array of stars, clusters, galaxies, and nebulae. The newer names for southern constellations often honor the tools of science and art. The iconic Southern Cross, or Crux, includes several notable stars as well as the "Jewel Box" open cluster and the Coalsack Nebula, a dense cloud of gas and dust silhouetted against the Milky Way. The southern sky also contains the huge constellation Centaurus—location of Alpha Centauri, neighbor to the sun and only 4.3 light-years away—as well as the Large and Small Magellanic Clouds, irregular dwarf galaxies that orbit the Milky Way. If you have an opportunity to visit the Southern Hemisphere, check out the wonders of its night sky. Many online resources can guide your tour.

||||||||||||||||||||||||||||||||||||||||||||||||||||||||||||||||||||||||||||

Zodiac

A group of star patterns recognized from ancient times, the 12 zodiac constellations represent animals, people, creatures—and one inanimate object, Libra—that appear in an annual cycle.

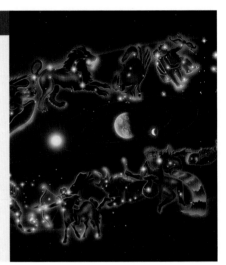

KEY FACTS

The zodiac constellations are Aries, Taurus, Gemini, Cancer, Leo, Virgo, Libra, Scorpius, Sagittarius, Capricornus, Aquarius, and Pisces.

+ fact: Several constellations of the zodiac are always in view.

+ fact: These constellations are located along the ecliptic, the sun's apparent yearly path across the sky.

The word "zodiac" comes from the Greek for "circle of animals," and most of the zodiacal constellations do have animal associations. In the sky, these constellations form an annual procession along the ecliptic, the apparent path taken by the sun across the sky as the Earth orbits the sun. As each of the 12 zodiacal constellations reaches its westernmost phase, it is in alignment with the sun, creating the idea of "sun sign" that forms the basis of the pseudo-science of astrology. Constellation Ophiuchus also reaches into the ecliptic path, but it usually is excluded from the zodiac. In Arab astronomy, even Libra (the Scales) had an animal connection, as reflected in the names of its alpha and beta stars, which were considered the claws of the constellation Scorpius.

Astrology

In the mystical and mind-bending 1970s, "What's your sign?" was a frequent conversation starter, reflecting the influence of astrology—at least superficially—in those changing times.

KEY FACTS

For millennia, astrology was considered a scholarly pursuit, and it remains so in some cultures.

+ fact: The Babylonians developed the first organized system of astronomy.

+ fact: Astrology helped promote the advancement of astronomy.

+ fact: Arab astrologers made many notable contributions to astronomy.

Few people are unaware of their "sign," the constellation of the zodiac where the sun resided when they were born. The correlation between signs and personal characteristics and the alignment of the sun, moon, stars, and planets at a particular point with future events forms the basis of astrology. Astrology is a practice as old as stargazing itself. The ancient Egyptians, for example, looked to the rising of the star Sirius to predict the flooding of the Nile, events that did coincide. Astrologers deem such occurrences to be cause-and-effect situations, using reliable observations to support their claims. In some cultures, predictions based on astrological interpretations are still taken into consideration when planning life events such as marriages and making other important decisions.

Constellation Charts

The following section of this chapter presents the night sky in the Northern Hemisphere through a series of four seasonal sky charts or maps and following those, through focused charts that highlight discussions of 56 different constellations.

The seasonal sky charts—one each for winter, spring, summer, and autumn—show the visible constellations, the brightest stars, and the locations of many galaxies, nebulae, and other deep sky objects. These maps reflect the fact that the celestial sphere operates on a constantly changing continuum.

Getting Oriented

These charts show the sky at a particular time but are useful any night. The maps are prepared from the perspective of an observer at latitude 40° N, at the dates and times indicated on each chart. This latitude line runs near Denver, New York, Madrid, and Beijing. Like the sun, the stars and constellations appear to move from east to west, in some cases falling from view below the horizon for several months. Observers at latitudes above 40° N will find the northern constellations higher in the sky, for longer periods of time, while losing sight of some below the southern horizon. The reverse is true for observers at positions toward the Equator.

The seasonal sky charts include a silhouetted landscape around the border indicating the horizon. Constellations appearing within about 10 degrees of that edge will be difficult to spot due to obstructions on Earth. The sky appears flat on the page, but arches in a hemisphere above the observer. The zenith is located in the center of the chart, the point directly overhead. Make sure you rotate the chart so that the word for the direction you are facing appears right side up. On the charts, stars with magnitudes down to 5 are connected by lines to form the constellations. Many stars of magnitude 3.5 or brighter (meaning a lower numerical value) are labeled for reference, and a star's size reflects its magnitude.

||||||||||||||||||||||||||||||||||||||||||||||||||||||||||||||||||||||||||||||||||||||||||||||||||

Constellation Charts

The pages that follow the seasonal sky charts describe 56 of the 88 recognized constellations. Many of the original 48 in Ptolemy's *Almagest,* an astronomical handbook, are included, as are newer constellations added to fill in the sky in the Northern Hemisphere. Those not included are mainly the "deep south" constellations that cannot be viewed from the mid-northern latitudes. You can learn about southern sky constellations online or in astronomy guides that often include such information.

Starry Signposts

The brighter stars in the larger constellations will provide the best beginning reference points. Typically, charts will use the Latin constellation name and a Greek letter that designates a constellation's component stars in order of magnitude. But not all "alpha" stars are created equal. The reason Orion stands out is that it includes three of the biggest stars in the sky—Rigel, Betelgeuse, and Bellatrix. Orion is also a good chance to observe star coloring because it has a conspicuous red star and blue star. Canis Major includes Sirius, the very brightest, as a reference point. Gemini has the bright stars Castor and Pollux for reference points. Most likely, these will be

BEST VIEWING TIMES		
Constellation	Star Chart	Months
Auriga	Winter	December / January
Boötes	Summer	June
Cancer	Spring	March / April
Capricornus	Summer	August / September
Cygnus	Summer	August / September
Lyra	Summer	July / August
Orion	Winter	January / February
Pegasus	Autumn	September / October
Sagittarius	Summer	July / August
Taurus	Winter	January / February

The brightest stars in the sky leave the boldest trails.

indicated in your chart by large circles. Look for the biggest, determine when those constellations will come into your field of view, and spend some time with them.

Star Hopping

Once you can identify starry signposts, you'll be prepared to star hop to fainter star patterns. The Andromeda group of stars, for example, provides stepping-stones to help spot the Andromeda galaxy, one of the closest neighbors to the Milky Way. It can be visible to the unaided eye on a dark moonless night. In the late

BRIGHTEST STARS IN THE NORTHERN SKY

Name	Constellation	Magnitude	Season
Sirius	Canis Major	-1.50	Winter
Arcturus	Boötes	-0.05	Summer
Vega	Lyra	0.03	Summer
Capella	Auriga	0.08	Winter
Rigel	Orion	0.12	Winter
Procyon	Canis Minor	0.40	Winter
Betelgeuse	Orion	0.42	Winter

northern summer sky, bright Alpha Andromedae forms one corner of the asterism the Square of Pegasus. It also marks the "tip" of the head of the Chained Lady—Andromeda herself, in constellation form. Gamma Andromedae and Beta Andromedae follow in line. From there, move up to the next reasonably bright star, and then to the one after that, and you'll be in the neighborhood of the galaxy, which looks like an oval, hazy patch about as wide as a fingertip.

Even some of the signs of the zodiac require proper orientation. Cancer, for example, has no star brighter than 4th magnitude. But it rests between Gemini and Leo, two show-offs with bright stars that can help locate the scuttling crab. The Big Dipper (in Ursa Major) is easy to spot and points to the North Star, Polaris. Polaris is not the brightest star in the sky, but it is the most stable. Useful for orientation, Polaris will stay fixed in the northern sky rather than shift with the seasons as other stars do.

HOW TO MEASURE DISTANCE

Distance across the night sky is typically measured in degrees. You can roughly measure degrees and distance using only your hands. Outstretch your arms and hold out your hands against the sky. At arm's length, a thumb covers approximately 2° of sky, a fist covers about 10°, and a fully outstretched hand about 20°. These approximations can be used to estimate the width and height of constellations. Each constellation entry in this book gives a width measurement using this system. Of course, hands come in all sizes, so these measurements are just approximations for average adult hands.

Key Facts

Each constellation heading shows the "Size on the sky," using a thumb length, closed fist, or outstretched hand or hands held at arm's length to indicate the constellation's approximate size in the sky. The Key Facts for the constellations include the number of brighter stars found in the constellation. "Best viewed" indicates the best months to hunt for the constellation. "Location" indicates the appropriate seasonal sky chart for reference. The constellation's alpha star also is noted.

Each constellation entry features a star chart showing the stars that make up a constellation and a portion of the surrounding sky.

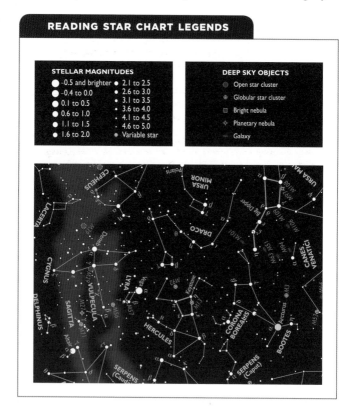

READING STAR CHART LEGENDS

STELLAR MAGNITUDES
- -0.5 and brighter
- -0.4 to 0.0
- 0.1 to 0.5
- 0.6 to 1.0
- 1.1 to 1.5
- 1.6 to 2.0
- 2.1 to 2.5
- 2.6 to 3.0
- 3.1 to 3.5
- 3.6 to 4.0
- 4.1 to 4.5
- 4.6 to 5.0
- Variable star

DEEP SKY OBJECTS
- Open star cluster
- Globular star cluster
- Bright nebula
- Planetary nebula
- Galaxy

|||||||||||||||||||||||||||||||||||||||||||||||||||||||||||||||||||||||||||||||||||

Light lines connect the brighter member stars of each constellation;
prominent asterisms are named. The brightest stars are labeled
with letters of the Greek alphabet, beginning with alpha. Some
of the brighter and well-known stars are labeled with their proper
names and, where appropriate, tinted to indicate the approximate
color in the sky. Background stars, neighboring constellations, and
nearby deep sky objects also appear on the maps.

NOTES ABOUT SKY CHARTS

+ Stars: The brightest stars have common names, like Antares,
which often relate to ancient mythology. But most bright stars that
form constellations are typically known by Johann Bayer's 1603
method of calling the brightest star in a constellation "alpha," the
second brightest "beta," and so on. For example, Deneb is the
common name for the brightest star in the constellation Cygnus
(the Swan), but the star is also known as Alpha Cygni, and is often
labeled with both names.

+ Deep sky objects: Most of the galaxies and other "fuzzy"
objects seen by backyard astronomers take their names from either
the Messier catalog (M) or the New General Catalog (NGC), where
they are assigned a number. Charles Messier published his catalog
of about 100 objects in 1771. The New General Catalog, with
7,840 objects, came in 1888. There are alternative names in these
catalogs and others. For example, in addition to the common name
Andromeda galaxy, the object is also known as M31 and NGC 224,
as well as by other names.

+ Mythological icons: These small drawings are designed to help
relate the shapes of the constellations to the figures they represent
in mythology. They are not necessarily oriented in the direction they
are typically seen in the night sky, and their arbitrary line patterns
do not always match what is shown on the detailed star charts on
the following pages.

Star maps often show line patterns to help illustrate the shape typically recognized as a constellation. These line patterns vary from
source to source. Choosing which stars to show as part of a constellation drawing is arbitrary. The International Astronomical Union
(IAU) defines constellations as areas of the night sky, not as strings
of specific stars. The IAU's well-defined technical boundaries are not
shown on these charts.

Winter

Orion makes a good orientation point in the winter sky. Bright star Betelgeuse appears on Orion's right shoulder and Rigel forms the left foot. At the sword, you'll find the Orion Nebula (M42).

STELLAR MAGNITUDES

- ● -0.5 and brighter
- ● -0.4 to 0.0
- ● 0.1 to 0.5
- ● 0.6 to 1.0
- ● 1.1 to 1.5
- ● 1.6 to 2.0
- ● 2.1 to 2.5
- ● 2.6 to 3.0
- ● 3.1 to 3.5
- ● 3.6 to 4.0
- ● 4.1 to 4.5
- ● 4.6 to 5.0
- ● Variable star

DATE	TIME
12/21	11 p.m.
1/21	9 p.m.
2/1	8 p.m.

DEEP SKY OBJECTS

- Open star cluster
- Globular star cluster
- Bright nebula
- Planetary nebula
- Galaxy

Spring

Spring is a good time to hunt for galaxies, as the sky offers a good view of elusive deep sky objects. Virgo is prominent, together with its dense collection of galaxies.

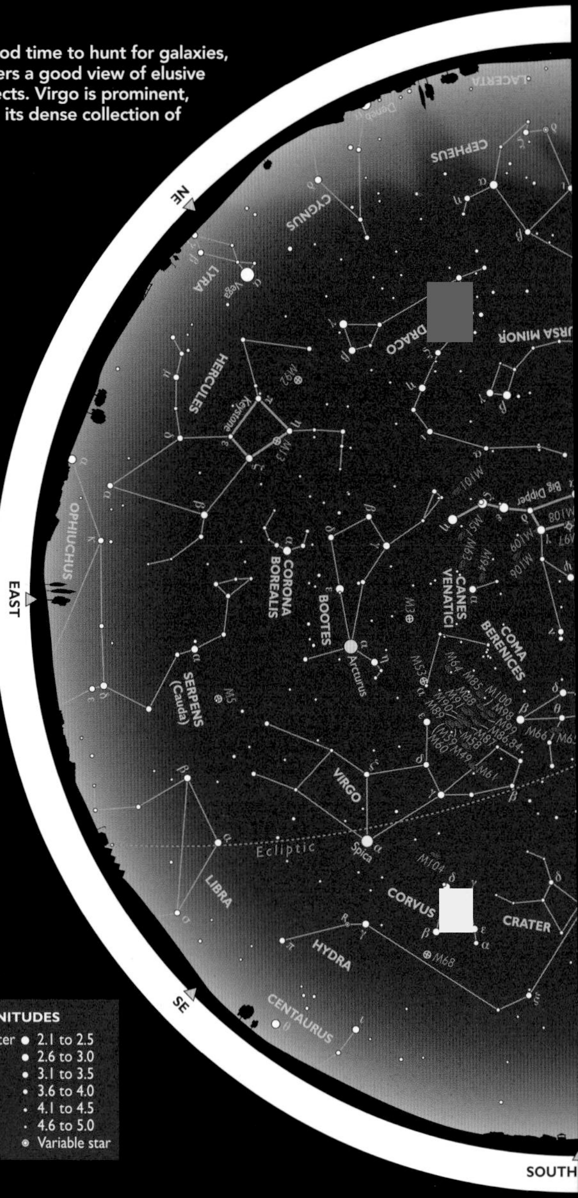

STELLAR MAGNITUDES

- ● -0.5 and brighter
- ● -0.4 to 0.0
- ● 0.1 to 0.5
- ● 0.6 to 1.0
- ● 1.1 to 1.5
- ● 1.6 to 2.0
- ● 2.1 to 2.5
- ● 2.6 to 3.0
- · 3.1 to 3.5
- · 3.6 to 4.0
- · 4.1 to 4.5
- · 4.6 to 5.0
- ● Variable star

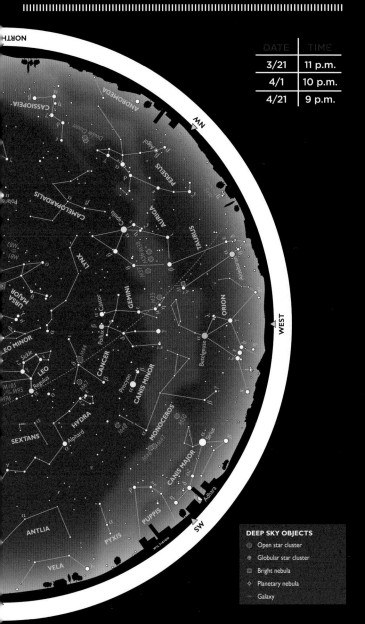

DATE	TIME
3/21	11 p.m.
4/1	10 p.m.
4/21	9 p.m.

DEEP SKY OBJECTS
- Open star cluster
- Globular star cluster
- Bright nebula
- Planetary nebula
- Galaxy

Summer

Summer places the Summer Triangle overhead,
highlighting three constellations: Lyra,
Cygnus, and Aquila. Vega will be almost
directly overhead, unmistakably
bright at magnitude 0.

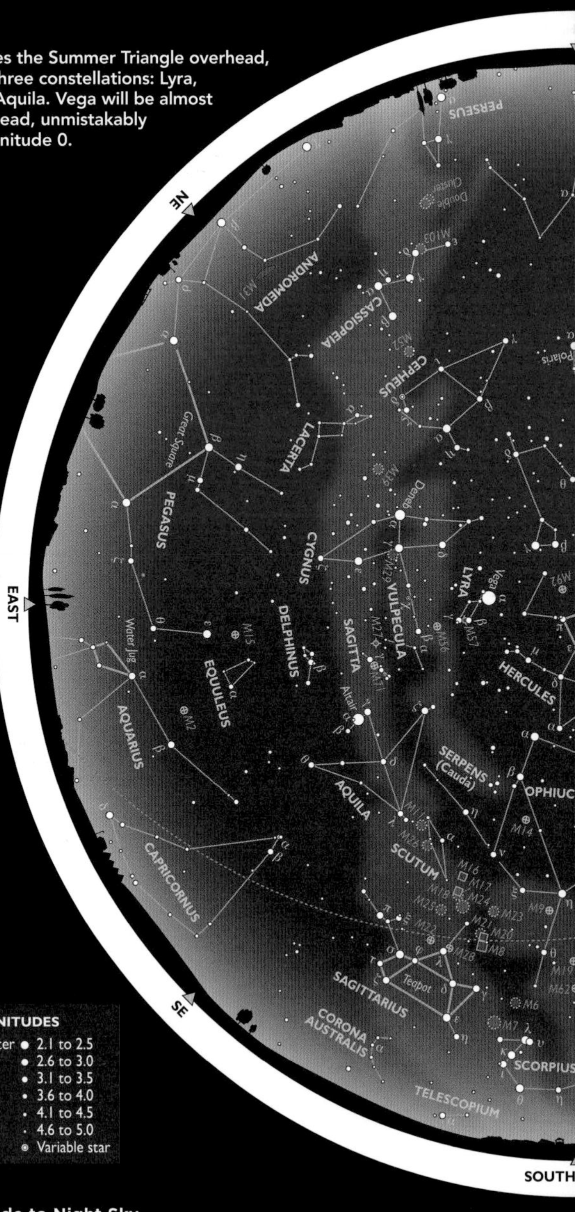

STELLAR MAGNITUDES

- -0.5 and brighter
- -0.4 to 0.0
- 0.1 to 0.5
- 0.6 to 1.0
- 1.1 to 1.5
- 1.6 to 2.0
- 2.1 to 2.5
- 2.6 to 3.0
- 3.1 to 3.5
- 3.6 to 4.0
- 4.1 to 4.5
- 4.6 to 5.0
- Variable star

DATE	TIME
6/21	11 p.m.
7/1	10 p.m.
7/21	9 p.m.

NORTH

AURIGA

CAMELOPARDALIS

LYNX

LEO MINOR

URSA MINOR

URSA MAJOR

M82 M81

M108 M97

M109

Big Dipper

LEO

Regulus

DRACO

M101

M51

M63

CANES VENATICI

COMA BERENICES

M106

NW

WEST

Keystone

CORONA BOREALIS

BOÖTES

Arcturus

M3

M64

M94

M53

M85 M100

M88 M91

M90 M89

M87 M58

M86 M84

M59 M60

M49

M61

VIRGO

β

SW

SERPENS (Caput)

M5

HERCULES

M12

M10

M107

M80

Antares

M4

LIBRA

Spica

Ecliptic

CORVUS

HYDRA

LUPUS

CENTAURUS

WESTERN

HORIZON

DEEP SKY OBJECTS

⚬ Open star cluster

⊕ Globular star cluster

▫ Bright nebula

✧ Planetary nebula

⬯ Galaxy

Autumn

The Great Square of Pegasus is central to locating several other autumn constellations, including Andromeda. The Andromeda galaxy (M31) is visible to the naked eye. Autumn also brings four major meteor showers.

STELLAR MAGNITUDES

- -0.5 and brighter
- -0.4 to 0.0
- 0.1 to 0.5
- 0.6 to 1.0
- 1.1 to 1.5
- 1.6 to 2.0
- 2.1 to 2.5
- 2.6 to 3.0
- 3.1 to 3.5
- 3.6 to 4.0
- 4.1 to 4.5
- 4.6 to 5.0
- Variable star

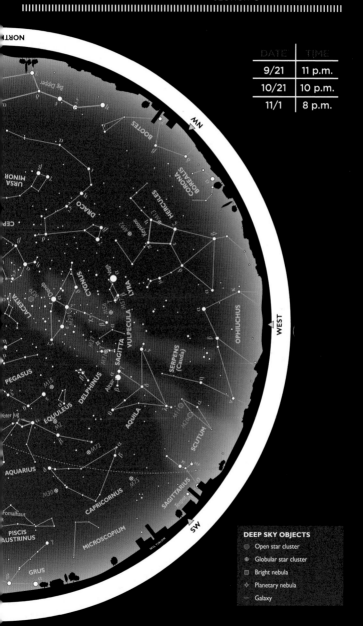

DATE	TIME
9/21	11 p.m.
10/21	10 p.m.
11/1	8 p.m.

NORTH

NW

WEST

SW

Big Dipper

URSA MINOR

BOÖTES

CORONA BOREALIS

DRACO

CEP

HERCULES

Keystone M92

M13

CYGNUS

LACERTA

Deneb

LYRA

M57

OPHIUCHUS

VULPECULA

SAGITTA

SERPENS (Cauda)

PEGASUS

M15

DELPHINUS

Altair

EQUULEUS

M2

AQUILA

SCUTUM

M11

Water Jug

M72

AQUARIUS

M30

SAGITTARIUS

CAPRICORNUS

M75

Fomalhaut

MICROSCOPIUM

PISCIS AUSTRINUS

WILTMON

GRUS

DEEP SKY OBJECTS

- Open star cluster
- Globular star cluster
- Bright nebula
- Planetary nebula
- Galaxy

Andromeda

The Chained Maiden   Size on the sky:

This constellation notably contains the Andromeda galaxy (M31), a spiral-shaped deep sky galaxy similar to our own Milky Way. Andromeda galaxy is visible to the naked eye.

KEY FACTS

The figure of Princess Andromeda, most often seen upside down, contains 7 main stars. Locate by tracing a line northeast from the northeast corner of the Great Square of Pegasus.

+ months best viewed: October–November

+ seasonal chart location: Autumn, center of chart

+ alpha star: Alpheratz

Cassiopeia and Cepheus, the mythical rulers of Ethiopia, angered the Greek gods when Cassiopeia boasted that her daughter Andromeda's beauty surpassed that of the daughters of Nereus, god of the sea and Poseidon's father-in-law. In retaliation, Poseidon sent the sea monster Cetus to destroy the kingdom. Cassiopeia and Cepheus then offered Andromeda as a sacrifice, chained to a rock. At the last moment, the hero Perseus, homeward bound with the head of Medusa, rescued Andromeda, his trophy head turning Cetus to stone in the bargain. Other characters in this famous tale, including Pegasus, the hero's mount, appear nearby in the sky.

|||||||||||||||||||||||||||||||||||||||||||||||||||||||||||||||||||||||||||||||||

Antlia
The Air Pump *Size on the sky:*

Northern Hemisphere stargazers can pick out this small
constellation in the spring near the southern horizon. It contains
two binary stars, one of which can be split with good binoculars.

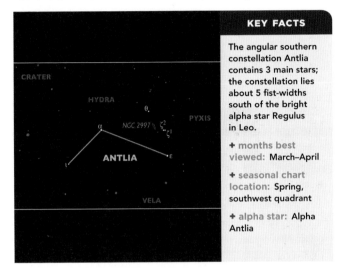

KEY FACTS

The angular southern
constellation Antlia
contains 3 main stars;
the constellation lies
about 5 fist-widths
south of the bright
alpha star Regulus
in Leo.

+ **months best
viewed:** March–April

+ **seasonal chart
location:** Spring,
southwest quadrant

+ **alpha star:** Alpha
Antlia

As a southern constellation, Antlia's backstory belongs
to the realm of science, not mythology. In the
1750s, French astronomer Nicolas-Louis de Lacaille
constructed the constellation from his vantage point at
the Cape of Good Hope in South Africa. He named it for
Irish chemist and physicist Robert Boyle's discovery of the
pneumatic pump, originally calling it Antlia Pneumatica.
All together, Lacaille charted some 10,000 southern stars.

Antlia's alpha star is its brightest one, but it has no proper
name. Antlia harbors the
galaxy NGC 2997, faintly visible
through a small telescope, just inside
its corner.

III

Aquarius

The Water Bearer   Size on the sky:

A constellation of the zodiac, Aquarius occupies a lot of celestial real estate along the ecliptic—the apparent path of the sun across the night sky—between Pisces and Capricornus.

KEY FACTS

Its 13 main stars form the spread-out figure of the Water Bearer; within this constellation, the Y-shaped Water Jug asterism is the most prominent feature.

+ **months best viewed:** September–October

+ **seasonal chart location:** Autumn, southwest quadrant

+ **alpha star:** Sadalmelik

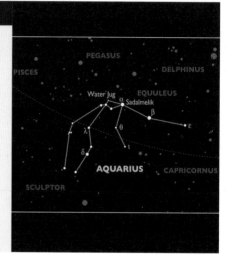

One of the fainter constellations of the zodiac, Aquarius is located south of the Great Square of Pegasus, in the "watery" section of the sky with associated constellations such as Cetus the Whale and Pisces. Aquarius has several mythological associations. The ancient Egyptians linked the constellation to the yearly flooding of the Nile River, the most vital event of the agricultural year, and the hieroglyph for water became the zodiacal symbol for Aquarius. To the Greeks, Aquarius was Zeus's cup-bearer, Ganymede. In some versions, he pours water or wine from a jug that maintains the celestial river Eridanus.

Aquila
The Eagle   Size on the sky:

Named by ancient Mesopotamian stargazers, Aquila the Eagle appears close enough to Earth's Equator to be seen from any terrestrial viewing position.

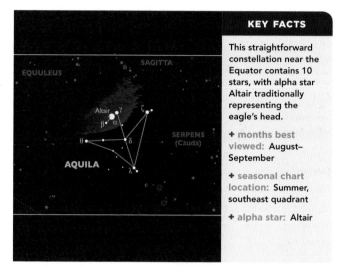

KEY FACTS

This straightforward constellation near the Equator contains 10 stars, with alpha star Altair traditionally representing the eagle's head.

+ **months best viewed:** August–September

+ **seasonal chart location:** Summer, southeast quadrant

+ **alpha star:** Altair

Aquila's alpha star, Altair, is one of the brightest stars in the sky and a good reference point for identifying the constellation. It also forms part of the Summer Triangle asterism along with Vega in constellation Lyra and Deneb in Cygnus. Aquila occurs in the Milky Way in an area of abundant star fields. In Greek myth, the eagle was Zeus's companion and carried the god's thunderbolts for him. He also is thought to have carried the young shepherd Ganymede to the sky to serve as Zeus's celestial cupbearer. Ganymede is immortalized nearby as the constellation Aquarius.

||||||||||||||||||||||||||||||||||||||||||||||||||||||||||||||||||||||||||||||||

Aries

The Ram   Size on the sky: 🖐

Ancient astronomers traditionally placed Aries at the beginning of the celestial zodiac because, millennia ago, the sun was "in" Aries at the time of the vernal equinox in spring.

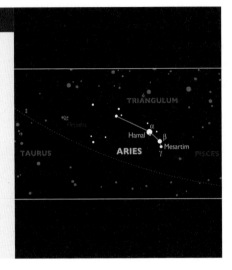

KEY FACTS

The brightest part of constellation Aries contains 4 stars; the "tail" of the ram is Gamma Arietis, or Mesartim, a double star with a wide separation.

+ months best viewed: November–December

+ seasonal chart location: Winter, southwest quadrant

+ alpha star: Hamal

On evenings in late fall and early winter, Aries appears high in the west between the Great Square of Pegasus and the Pleiades asterism in Taurus. The constellation's double star, Mesartim, was one of the first doubles spotted with a telescope—by astronomer Robert Hooke in 1664. In the view of the ancient Greeks, Aries was the source of the Golden Fleece stolen by Jason and the Argonauts. Jason wrested the fleece from the custody of a dragon that had received it from a king. The king had sacrificed the ram in gratitude for saving his two children from an abusive stepmother.

||||||||||||||||||||||||||||||||||||||||||||||||||||||||||||||||||||||||||||||

Auriga

The Charioteer *Size on the sky:*

The tidy, elegant form of Auriga appears in the heart of the Milky Way and is on Ptolemy's list of the 48 original ancient constellations.

KEY FACTS

The constellation Auriga, found on a line between Orion and Polaris, has 7 main stars. It shares with Taurus a star (lowest on the map) that otherwise would be Auriga's gamma star.

+ months best viewed: December–January

+ seasonal chart location: Winter, center of chart

+ alpha star: Capella

Auriga's shining glory is its alpha star, the brilliant Capella, the sixth brightest star in the sky. Southwest of Capella lies Epsilon Aurigae, an eclipsing binary star veiled every 27 years by a companion star. The constellation lies along the galactic equator, the great circle that passes through the densest part of the Milky Way and contains several interesting star clusters visible through binoculars, notably M36 and M37. The chariot and rider of Auriga may represent Hephaestus of Greek myth, the crippled blacksmith god who built the chariot to accommodate his handicap.

Boötes

The Herdsman *Size on the sky:*

The ancient constellation of Boötes is a celestial highlight of early summer. Its alpha star, Arcturus, ranks as fourth brightest star in the sky.

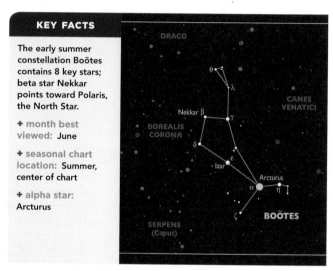

KEY FACTS

The early summer constellation Boötes contains 8 key stars; beta star Nekkar points toward Polaris, the North Star.

+ **month best viewed:** June

+ **seasonal chart location:** Summer, center of chart

+ **alpha star:** Arcturus

Boötes's alpha star, Arcturus, lies on an arc continuing from the handle of the Big Dipper, giving rise to the mnemonic "arc to Arcturus." In a variation of the myth surrounding the constellation, herdsman Boötes keeps his flock moving about the sky in pursuit of Ursa Major and Ursa Minor. In another, he herds a bear; a loose translation of Arcturus is "bear keeper." In the first week in January, the northern part of Boötes hosts the Quadrantid meteor shower, one of the strongest of the year, producing several dozen meteors an hour during its peak. It occurs at the point where Boötes, Hercules, and Draco meet.

||||||||||||||||||||||||||||||||||||||||||||||||||||||||||||||||||||||||||||||||||||||

Camelopardalis
The Giraffe   Size on the sky:

Camelopardalis is located in a region of the sky where few stars are visible. No star in this faint constellation has a common name.

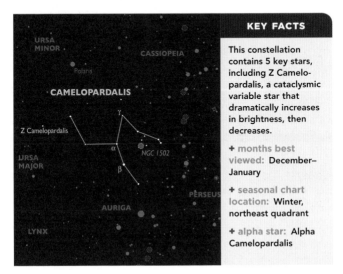

KEY FACTS

This constellation contains 5 key stars, including Z Camelopardalis, a cataclysmic variable star that dramatically increases in brightness, then decreases.

+ **months best viewed:** December–January

+ **seasonal chart location:** Winter, northeast quadrant

+ **alpha star:** Alpha Camelopardalis

Camelopardalis is a modern constellation, created in 1613 to fill the celestial gap between the bears, Ursa Major and Ursa Minor, and Perseus. It neighbors Polaris, the North Star, making it a strictly northern constellation, although a faint one. It contains the star cluster NGC 1502, a good target for a telescope, and is near spiral galaxy NGC 2403, which lies about 12 million light-years from Earth. Although the ancient Greeks called the giraffe a "camel-leopard" (and gave the animal its species name: *camelopardalis*), the constellation possibly was named for the biblical camel that carried Rebecca to her marriage with Isaac.

Cancer

The Crab   Size on the sky:

As a constellation of the zodiac, tiny Cancer the Crab is far less noticeable than most. It has no stars with a magnitude brighter than 4.

KEY FACTS

The 5 dim stars of the tiny, faint constellation Cancer contain the Beehive cluster (M44), which is visible to the naked eye under dark skies.

+ months best viewed: March–April

+ seasonal chart location: Spring, southwest quadrant

+ alpha star: Acubens

LEO

CANCER

γ Asellus Borealis
Asellus Australis   M44
                     (Beehive Cluster)
δ
Acubens
α
M67

β

HYDRA

Unassuming Cancer contains some interesting star groups. Viewed with binoculars, the Beehive cluster reveals more than 20 stars, while open cluster M67, with its 500 stars, can be seen through a small telescope. The Tropic of Cancer gets its name from the constellation, although the sun no longer is in it on the summer solstice as it was for early mapmakers. Cancer's story involves Zeus's often jealous wife, Hera. She sent the crab to thwart Hercules, Zeus's son with a mortal, as he battled the multiheaded monster Hydra. Proving no match, the crab perished under the hero's foot, but earned a place in the stars.

Canes Venatici

The Hunting Dogs   Size on the sky:

Location is everything in the night sky: The two stars of Canes Venatici appear just about where two leashed hounds would be expected in relation to Boötes, the Herdsman.

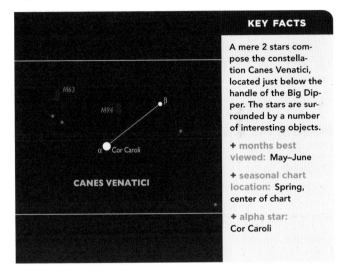

KEY FACTS

A mere 2 stars compose the constellation Canes Venatici, located just below the handle of the Big Dipper. The stars are surrounded by a number of interesting objects.

+ months best viewed: May–June

+ seasonal chart location: Spring, center of chart

+ alpha star: Cor Caroli

An alternative way of viewing Canes Venatici is to look for the two stars running between the legs of Ursa Major, the bear that the dogs are chasing. The small northern constellation was created by Polish astronomer Johannes Hevelius in the 17th century and has been accepted as one of the official 88 constellations. The name of its alpha star, Cor Caroli, stands for the Heart of Charles and was reputedly bestowed on the star by famous English astronomer Edmond Halley in honor of his sponsor, King Charles II. The 500,000-star globular cluster M3 can be found midway between Cor Caroli and Arcturus in Boötes.

Canis Major

The Larger Dog   Size on the sky:

Canis Major appears just to the southeast of the constellation Orion, not far from the Milky Way. The constellation's alpha star, Sirius, outshines all other stars in the night sky.

KEY FACTS

Canis Major contains 8 key stars, most notably Sirius, brightest in the night sky; smaller stars, star clusters, and nebulae crowd around and within the constellation.

+ months best viewed: January–February

+ seasonal chart location: Winter, southeast quadrant

+ alpha star: Sirius

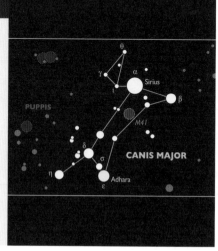

Sirius, the Dog Star, shows the way to Canis Major, the constellation representing the larger of Orion's hunting dogs. A line traced through Orion's belt and continuing southeast points directly to Sirius within its constellation. The Dog Star figures in the expression "dog days of summer." In late summer, Sirius rises around the same time as the sun, and the two "conspire" to generate extra warmth in the Northern Hemisphere. Several star clusters and nebulae are located in Canis Major. The brightest, open cluster M41, is easily spotted through binoculars and even more impressive through the lens of a telescope.

Canis Minor

The Smaller Dog   Size on the sky:

Orion's smaller hunting dog, Canis Minor, lies directly northeast of its larger companion, Canis Major. The dim constellation lacks any bright deep sky objects.

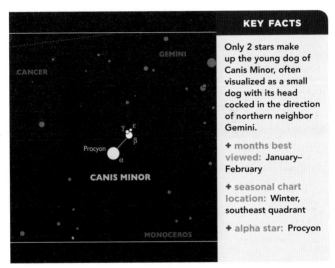

KEY FACTS

Only 2 stars make up the young dog of Canis Minor, often visualized as a small dog with its head cocked in the direction of northern neighbor Gemini.

+ **months best viewed:** January–February

+ **seasonal chart location:** Winter, southeast quadrant

+ **alpha star:** Procyon

Multiple scenarios explain the dim and subdued Canis Minor. One puts the dog beneath the table of Castor and Pollux, the Gemini twins, waiting for scraps. Another makes him Helen of Troy's favorite pup that allowed her to elope with the Trojan prince Paris. Together with Betelgeuse in Orion and Canis Major's Sirius, Canis Minor's alpha star, Procyon, forms the Winter Triangle, a seasonal asterism that helps orient winter stargazers. Procyon lies only 11.2 light-years from Earth and is the sixth brightest star in the northern night sky. The constellation also hosts the Canis Minorids, a faint meteor shower of early December.

Capricornus
The Sea Goat   Size on the sky:

The broad and distinctive triangle of Capricornus stands out in
the southern sky, especially in late summer and early fall. Algedi,
Arabic for "the goat," is the zodiacal constellation's alpha star.

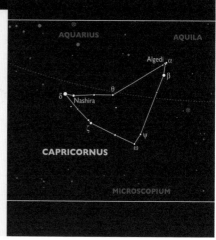

KEY FACTS

The 12 stars of Cap-
ricornus are not very
bright, but their
recognizable shape
trumps magnitude.
They lie southeast of
the bright star Altair in
Aquila.

**+ months best
viewed:** August–
September

**+ seasonal chart
location: Summer,
southeast quadrant**

+ alpha star: Algedi

Long recognized as a goat, Capricornus at some point
acquired a fish tail. In one story, the god Pan, a satyr,
leaped into the River Nile to escape a monster, and the
water transformed him. In an older tale, the animal represents one of Zeus's warriors, discoverer of conch shells
with a resounding call that frightened the
opposing Titans into retreat. Zeus placed
the warrior in the sky with a fish tail and
horns to represent his discovery. Alpha star
Algedi appears to the naked eye
as an optical binary composed of
two stars that are closely aligned
but separated by more than 500
light-years.

II

Cassiopeia

The Queen   Size on the sky:

Visible year-round because of its proximity to the north celestial pole, Cassiopeia's iconic W shape (or M, depending on the season) makes identification easy.

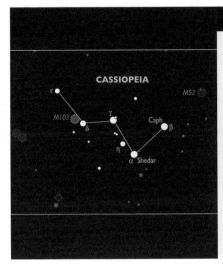

CASSIOPEIA

ε
M103
δ
γ
η
α Shedar
Caph
β
M52

KEY FACTS

The 5 stars of this iconic constellation form the figure of Queen Cassiopeia on her throne within a portion of the Milky Way facing Polaris, the North Star; visible much of year.

+ months best viewed: October–November

+ seasonal chart location: Autumn, northeast quadrant

+ alpha star: Shedar

Chained to her throne, Ethiopian Queen Cassiopeia sits between the constellations of her husband, Cepheus, and her daughter, Andromeda. Boasting of her daughter's good looks, the queen drew the ire of the sea god Poseidon, who felt that his own daughters, the Nereids, had been disrespected. Cassiopeia and Cepheus offered to sacrifice Andromeda to save their kingdom. At the last moment, Perseus saved Andromeda, but Cassiopeia was punished nonetheless, forced to hang upside down half the year. Visible through a telescope, the dense open cluster M52, seen off the leg of the W, contains about 100 stars.

Cepheus

The King   Size on the sky:

A circumpolar constellation, Cepheus is visible all year in the Northern Hemisphere, but the brightness of its main stars competes with surrounding stars, often making it hard to spot.

KEY FACTS

The body of King Cepheus, facing the open end of Cassiopeia, contains 5 key bright stars. They take the shape of a small house pointing generally toward Polaris, the North Star.

+ months best viewed: September–October

+ seasonal chart location: Autumn, center of chart

+ alpha star: Alderamin

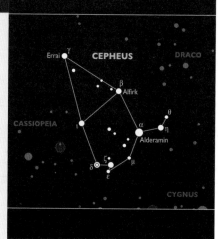

King Cepheus, the hapless consort of Queen Cassiopeia, was forced to put daughter Andromeda at peril because of his wife's unwise boast of their daughter's beauty, which angered the sea god Poseidon. As in his mythological life, the constellation Cepheus plays second fiddle to that of his wife, facing the open end of the W shape of the constellation Cassiopeia. Errai, Cepheus's gamma star, is both a binary star and host to an orbiting planet. Estimated to be almost 1.6 times the size of Jupiter, Errai's planet attests that planets can form in relatively close binary systems, stars that orbit the same center of gravity.

Cetus

The Sea Monster   *Size on the sky:*

Very large and faint, the constellation Cetus appears in a sky neighborhood known as the Heavenly Waters in the company of constellations Eridanus and Pisces.

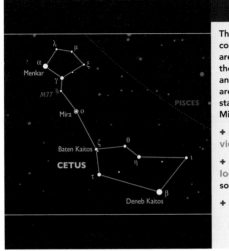

KEY FACTS

The 13 stars of the constellation Cetus are divided between the monster's head and body. The parts are connected by two stars; one of them, Mira, is a variable star.

+ **month best viewed:** November

+ **seasonal chart location:** Autumn, southeast quadrant

+ **alpha star:** Menkar

The god Poseidon sent the frightful sea monster Cetus—who gave his name to the mammalian order Cetacea, the whales—to terrorize Ethiopia after Queen Cassiopeia insulted Poseidon's daughters, the Nereids, by boasting that her daughter was more beautiful. Cetus was vanquished by Perseus and turned to stone by the head of Medusa.

The constellation is also associated with the whale that swallowed the prophet Jonah in the Old Testament. In autumn, the head of Cetus appears between Taurus and Pisces in the southern sky, with its body bordering Aquarius.

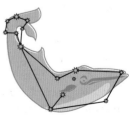

Columba

The Dove   Size on the sky:

Columba is a smallish constellation, one of the modern ones added in the 16th century to fill out the sky chart. It honors the dove Noah sent out from the ark to search for dry land.

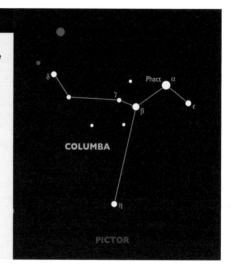

KEY FACTS

The smallish constellation Columba, made up of 8 stars, has the appearance of three legs spiraling from a medium-bright star at the center. Nearby sky objects often outshine it.

+ **months best viewed:** January–February

+ **seasonal chart location:** Winter, southeast quadrant

+ **alpha star:** Phact

A t the end of the 40 days and nights of epic rain that flooded all the Earth, Noah sent out a dove to see if the waters of the great flood had begun to recede. When the dove returned carrying an olive sprig in its bill, Noah knew immediately that it had. The dove's association with the ark and the flood made it a natural companion of the watery constellations of the Heavenly Waters group. The constellation is located very close to brighter objects in the night sky, making the Dove somewhat difficult to make out, especially for novice sky-watchers.

III

Coma Berenices

Berenice's Hair   Size on the sky:

Coma Berenices formed part of other constellations—a wisp of Virgo's hair or tuft of Leo's tail—until 17th-century astronomer Tycho Brahe helped designate it a constellation in its own right.

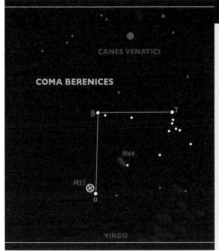

KEY FACTS

Located just north of Virgo and south of Canes Venatici near the Virgo cluster of galaxies, the rough right angle that is Coma Berenices contains 3 stars.

+ months best viewed: **May–June**

+ seasonal chart location: **Spring, southeast quadrant**

+ alpha star: **Alpha Comae Berenices**

A rejuvenated tale of an Egyptian queen provides a namesake and story for Coma Berenices. The consort of Ptolemy III, Berenice sought Aphrodite's aid in the safe return of her husband from war, promising the goddess luxurious hair as an incentive. After the wish was granted, an astronomer of the royal court convinced the ruling couple that a grateful Aphrodite had placed the queen's gift in the stars. The constellation contains the deep sky spiral galaxy M64, known as the Black Eye. It lies between the constellation's two outermost stars, along the line that would represent the base of a triangle.

|||||||||||||||||||||||||||||||||||||||||||||||||||||||||||||||||||||||||||||||||||||||||

Corona Australis

The Southern Crown Size on the sky: ✋

Though an original Ptolemaic constellation, Corona Australis is primarily southern, rising only a few degrees above the horizon during midsummer in the mid-northern latitudes.

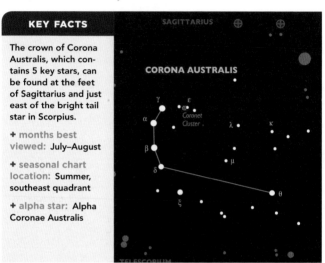

KEY FACTS

The crown of Corona Australis, which contains 5 key stars, can be found at the feet of Sagittarius and just east of the bright tail star in Scorpius.

+ months best viewed: July–August

+ seasonal chart location: Summer, southeast quadrant

+ alpha star: Alpha Coronae Australis

For a small, faint constellation, Corona Australis enjoys a wealth of supporting mythology, much in the form of stories referring to a crown of laurel or fig leaves. One version regards it as the crown of Chiron, the Centaur. Another puts Apollo in the picture, having him fashion the crown from the leaves of his love Daphne, who was changed into a laurel tree to escape the insistent Apollo's advances. The constellation hosts an active star-forming region composed of the Coronet cluster and the nearby Corona Australis Nebula. Though visible only through advanced deep sky technology, this region boasts about 30 "newborn" stars.

||||||||||||||||||||||||||||||||||||||||||||||||||||||||||||||||||||||||||||||

Corona Borealis
The Northern Crown *Size on the sky:* 👆

Despite having one of the more definitive constellation shapes, small and faint Corona Borealis is squeezed in between its bigger, brighter neighbors Boötes and Hercules.

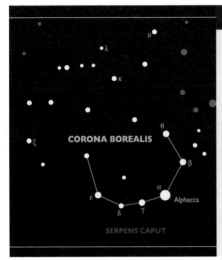

μ
λ
κ
CORONA BOREALIS
θ
ξ
β
α
Alphecca
ε
δ γ
SERPENS CAPUT

KEY FACTS

An obvious, semicircular crown shape made of 7 stars in a small and faint constellation is located just south of a line traced between the bright stars Vega in Lyra and Arcturus in Boötes.

+ months best viewed: June–July

+ seasonal chart location: Summer, center of chart

+ alpha star: Alphecca

Another Greek crown-related myth provides a plausible tale for the Northern Crown. In it, good-time god Dionysus threw his crown into the sky to impress Ariadne, princess of Crete and his future wife. She had rebuffed him earlier when he courted in the form of a young mortal. The crown toss changed her mind and they married. In American Indian lore, this celestial shape represents a camp circle. Some of the stars in Corona Borealis vary significantly in brightness. One of them is being investigated as possibly being an extrasolar planet; a body larger than Jupiter already has been found.

||||||||||||||||||||||||||||||||||||||||||||||||||||||||||||||||||||||||||||||||||||||

Corvus

The Crow   Size on the sky:

Corvus, along with Crater, perch in a crook of Hydra, the serpentine megaconstellation. The Crow is located just west of Spica in nearby Virgo.

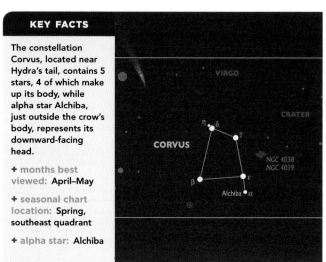

KEY FACTS

The constellation Corvus, located near Hydra's tail, contains 5 stars, 4 of which make up its body, while alpha star Alchiba, just outside the crow's body, represents its downward-facing head.

+ months best viewed: **April–May**

+ seasonal chart location: **Spring, southeast quadrant**

+ alpha star: **Alchiba**

In a myth belying the smarts of members of the Corvid family, a crow is sent for a cup of water by the god Apollo. The bird spied a tempting unripe fig and waited for it to ripen. Anticipating the god's ire, the crow made up a tale about being attacked by a snake, which he carried back along with the cup of water. Apollo saw through the deceit and banished the cup, the crow, and the serpent to the sky. A pair of colliding galaxies known as the Ring-tailed galaxy (NGC 4038 and NGC 4039) can be seen with an 8-inch-aperture telescope just outside Corvus, with the "tail" near the border of Corvus and Crater.

||||||||||||||||||||||||||||||||||||||||||||||||||||||||||||||||||||||||||||||||||||||||||||||||

Crater
The Cup   Size on the sky: 🖐

The small constellation Crater appears near its mythological associates, Hydra and Corvus. To some, Crater first represented a spike on sea monster Hydra's back.

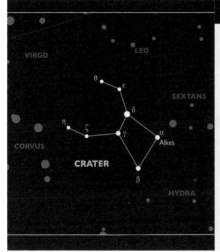

KEY FACTS

The 8 stars of the constellation Crater make a goblet shape: 4 stars form the base of the goblet and 4 more form the cup, which opens toward the star Spica in the constellation Virgo.

+ **months best viewed:** April–May

+ **seasonal chart location:** Spring, southeast quadrant

+ **alpha star:** Alkes

The unmistakable goblet shape of Crater can be found in the spring sky right above Hydra and just west of Corvus, an associated constellation. Crater represents the vessel brought to Apollo by Corvus, the crow. Apollo tossed the two into the sky, along with Hydra, when he realized Corvus had lied about the reason he was late in bringing a requested cup of water. Some sky observers also relate the cup to the constellation Aquarius, the Water Bearer, which occupies the same quadrant of the sky in summer. More recent astronomers associate the cup with the Holy Grail or Noah's wine goblet.

Cygnus

The Swan   Size on the sky:

Cygnus appears in a dense and visually rewarding portion of the sky. It is sometimes known as the Northern Cross for its shape and bright stars.

KEY FACTS

The 13 stars of the constellation Cygnus form recognizable wings and a neck and head that in late summer and fall point southward like a migratory bird.

+ months best viewed: August–September

+ seasonal chart location: Summer, northeast quadrant

+ alpha star: Deneb

The ancients associated several bird-centric myths with this prominent constellation. It was seen as Zeus, transforming himself into a swan to seduce Leda, and also as Orpheus, who was murdered and turned into a swan for rejecting a group of maidens. He was placed in the sky next to his beloved lyre, the constellation Lyra. Cygnus also represented one of the Stymphalian birds, the quarry Hercules pursued in one of the 12 labors. Alpha star Deneb, along with Altair and Vega, create the Summer Triangle, and a large diffuse gas emission nebula shaped like and named for North America appears on the Swan's southern border.

Delphinus

The Dolphin   Size on the sky:

The distinct shape of Delphinus, one of the sea creature constellations of the Heavenly Waters sky region, seems to swim toward Pegasus.

KEY FACTS

The constellation Delphinus contains 5 dim stars, 4 of which form the body, which is an asterism known as Job's Coffin; the fifth is the creature's curved tail.

+ **months best viewed:** August–September

+ **seasonal chart location:** Summer, southeast quadrant

+ **alpha star:** Sualocin

The constellation Delphinus holds its place in the sky because he was a favorite of the Greek sea god Poseidon. The little dolphin succeeded in convincing the Nereid Amphitrite, the sea nymph daughter of Nereus, to marry Poseidon after the god himself had attempted and failed to win her attention. Delphinus can be located just east of a straight line traced between Altair in the constellation Aquila and Deneb in the constellation Cygnus. The Dolphin constellation's gamma star is an optical double star, best viewed through a telescope, whose dimmer star has a greenish tinge.

||||||||||||||||||||||||||||||||||||||||||||||||||||||||||||||||||||||||||||||||||

Draco

The Dragon   Size on the sky:

Visible year-round in the Northern Hemisphere, Draco is one of the constellations closest to the north celestial pole. Draco contains the Cat's Eye Nebula (NGC 6543), a dying sunlike star.

KEY FACTS

Draco contains 18 stars; its dragon tail emerges between the bears Ursa Major and Ursa Minor and dips toward sea monster Cepheus, while its head points to Hercules.

+ months best viewed: **May–June**

+ seasonal chart location: **Spring, northeast quadrant**

+ alpha star: **Thuban**

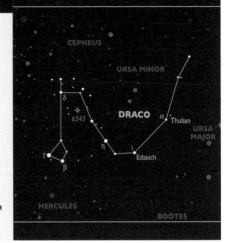

The constellation Draco represented different scaly beasts to different civilizations. To the ancient Greeks, it was the dragon Ladon, who was slain by Hercules. To the Hindus, it was a celestial alligator, and to the Persians, a giant serpent. Draco's alpha star, Thuban, once was the Earth's polestar, before the phenomenon of precession shifted the celestial position of Earth's axis and made Polaris the polestar. The Quadrantid meteor shower erupts from the region where Draco, Boötes, and Hercules meet in the beginning of January. One of the heaviest meteor showers of the year, the Quadrantids last only a few hours.

|||||||||||||||||||||||||||||||||||||||||||||||||||||||||||||||||||||||||||||||||||

Equuleus
The Little Horse   Size on the sky:

The second smallest constellation in the area, Equuleus resides in a crowded southern portion of the sky, where it is overshadowed by the larger horse constellation, Pegasus.

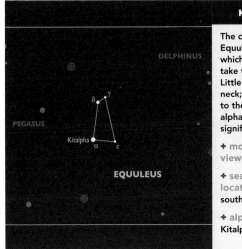

DELPHINUS

PEGASUS

δ γ

Kitalpha α ε

EQUULEUS

KEY FACTS

The constellation Equuleus has 4 stars, which essentially take the form of the Little Horse's head and neck; the body is left to the imagination. Its alpha star is the only significant feature.

+ months best viewed: July–August

+ seasonal chart location: Summer, southeast quadrant

+ alpha star: Kitalpha

Equuleus is the opening act for Pegasus, rising and setting before its neighbor, a fact that gave it the designation of Equus Prior. The constellation is thought to represent Celeris, the brother-horse of Pegasus. Pegasus is best known as the noble winged steed of Perseus, slayer of Medusa and savior of Andromeda. Celeris belonged to Castor, who along with twin Pollux represents the twins of Gemini. Castor was famed as a skilled equestrian and received the gift of Celeris from the messenger god Hermes. Look for Equuleus between neighbors Pegasus to the northeast and Aquila to the southwest.

Eridanus

The River   Size on the sky:

On a clear night in winter with an open horizon, you have a good chance of seeing Eridanus, although its southern tip disappears below the horizon at about 20° N latitude.

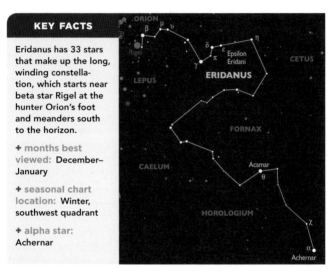

KEY FACTS

Eridanus has 33 stars that make up the long, winding constellation, which starts near beta star Rigel at the hunter Orion's foot and meanders south to the horizon.

+ **months best viewed:** December–January

+ **seasonal chart location:** Winter, southwest quadrant

+ **alpha star:** Achernar

The river of the constellation Eridanus was variously identified as the Euphrates or the Nile. The ancient Greeks saw it as the river into which Phaëthon, son of the sun god Helios, was cast after he failed to control the sun god's chariot and Earth stood in danger of burning up. It is a long river, to be sure, the sixth largest constellation in the sky containing its sixth brightest star, alpha star Achernar. Eridanus bends tightly at its beginning near Rigel and then gradually widens out. It contains a bright star similar to our sun, Epsilon Eridani, which is known to host a confirmed planet.

|||||||||||||||||||||||||||||||||||||||||||||||||||||||||||||||||||||||||||||||||||||||

Gemini
The Twins   Size on the sky:

The twins Castor and Pollux stride hand in hand in the southeast winter sky, northeast of the bright star Betelgeuse on constellation Orion's upraised arm.

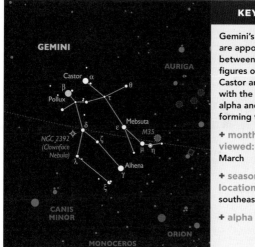

KEY FACTS

Gemini's 13 key stars are apportioned between the two stick figures of the twins Castor and Pollux, with the constellation's alpha and beta stars forming the heads.

+ months best viewed: February–March

+ seasonal chart location: Winter, southeast quadrant

+ alpha star: Castor

The Gemini twins represent Castor and Pollux, sons of the unions of Leda with a swan-disguised, seductive Zeus and her husband Tyndareus, and brothers of Helen of Troy and Clytemnestra—the exact paternity details remaining murky. Castor and Pollux also served as shipmates with Jason on the *Argo*. The impressive Geminid meteor showers originate from the constellation in the middle of December. Binoculars or a telescope yield views of M35, an open star cluster of hundreds of stars that resides near the three "foot stars" of Castor. The constellation also contains the blue-green Clownface Nebula, visible only through a telescope.

Grus
The Crane   Size on the sky:

Grus is a breakout constellation, which was separated from the formation Piscis Austrinus, the Southern Fish. It is visible in the Northern Hemisphere only briefly in the fall.

KEY FACTS

Grus has 11 stars that make up the rough form of a crane in the constellation, including the distinct central shape of an X, which is composed of brighter stars.

+ **month best viewed:** September

+ **seasonal chart location:** Autumn, southwest quadrant

+ **alpha star:** Alnair

The largely southern constellation Grus is a modern invention, named in 1603 by German celestial cartographer Johann Bayer in honor of the bird the ancient Egyptians made the symbol of the astronomer. Alternate names for the star group include Anastomas (the Stork), den Reygher (the Heron), and Phoenicopterus (the Flamingo). The constellation harbors the Grus Quartet, a group of four spiral galaxies (NGC 7552, 7582, 7590, and 7599) that are located very close together and exert strong influences on each other. Millions of years from now, the four galaxies will likely unite.

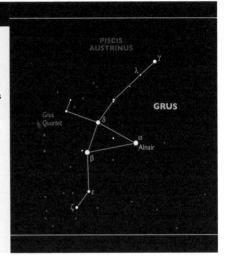

|||||||||||||||||||||||||||||||||||||||||||||||||||||||||||||||||||||||||||||||||||||||||

Hercules

The Superhero   Size on the sky:

The constellation Hercules contains a he-man's worth of stars, although none of them are exceptionally bright. Look for him overhead in the summer months.

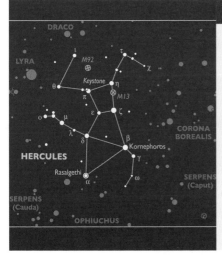

KEY FACTS

It takes 20 key stars to make up the hefty form of Hercules, including the recognizable backwards K at the center; his "torso" contains the 4-star asterism known as the Keystone.

+ **months best viewed:** July–August

+ **seasonal chart location:** Summer, center of chart

+ **alpha star:** Rasalgethi

The half-mortal son of Zeus, Hercules was plagued throughout much of his life by the jealousy of Hera, Zeus's wife and queen of the gods, who resented the tangible reminder of her husband's dalliance. She induced a fit of insanity in Hercules that caused him to murder his wife and children. In grief and repentance, he undertook 12 nearly impossible labors, and at their successful completion was granted immortality. The constellation contains M13, considered the best globular cluster visible in the northern sky. It occurs on one side of the Keystone asterism and can be seen as a blur by the naked eye.

||||||||||||||||||||||||||||||||||||||||||||||||||||||||||||||||||||||||||||||||||||||

Hydra

The Sea Serpent   *Size on the sky:*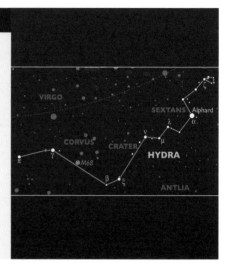

Hydra winds across the sky from Cancer to Libra and spans
the largest area of any constellation. It used to include today's
constellations Sextans, Corvus, and Crater.

KEY FACTS

The constellation
Hydra's 17 stars begin
with a kite-shaped
head, continue past
the "heart" at alpha
star Alphard, and end
with the sea serpent's
tail close to the south-
ern horizon.

**+ months best
viewed:** March–April

**+ seasonal chart
location:** Spring,
southwest quadrant

+ alpha star:
Alphard

In classical mythology, Hydra lost a decisive battle with
Hercules, who fought the multiheaded sea serpent as
one of his 12 labors. Hydra put up a good fight, regrow-
ing two heads for each one that Hercules lopped off, and
it looked pretty hopeless for Hercules until the stumps
of each decapitated head were cauterized to stop the
regrowth. Constellation Hydra's enormous span belies the
relative dimness of its stars; only two of them surpass a
magnitude of 3. Alphard, the brightest star in the formation, can be located by looking just east of a line traced
from Regulus in the
constellation Leo
to the star Sirius in
Canis Major.

|||||||||||||||||||||||||||||||||||||||||||||||||||||||||||||||||||||||||||||||||||||||

Lacerta

The Lizard   Size on the sky:

A so-called modern constellation, Lacerta was created by 17th-century Polish astronomer Johannes Hevelius. At first depicted as a small mammal, it later morphed into a lizard.

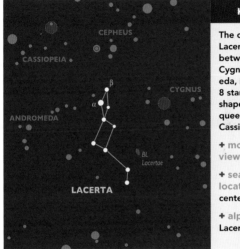

CEPHEUS

CASSIOPEIA

β

α

CYGNUS

ANDROMEDA

BL
Lacertae

LACERTA

KEY FACTS

The constellation Lacerta, which zigzags between Cassiopeia, Cygnus, and Andromeda, is composed of 8 stars. It forms a W shape like its brighter, queenly neighbor Cassiopeia.

+ month best viewed: **October**

+ seasonal chart location: **Autumn, center of the chart**

+ alpha star: **Alpha Lacertae**

Lacerta is a constellation without a relevant mythology or backstory. At the time it was created to fill in a gap on the celestial map, competing proposals were offered to name it for Louis XIV or Frederick the Great. Although Lacerta's stars are faint, they remain visible to observers for much of the year and appear almost directly overhead in early fall for those in mid-northern latitudes. Lacerta contains an intriguing deep sky object, BL Lacertae, which is the center of an elliptical galaxy visible only through a substantial telescope. Of great interest, its core may harbor a black hole.

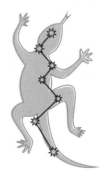

‖‖‖

Leo

The Lion   Size on the sky: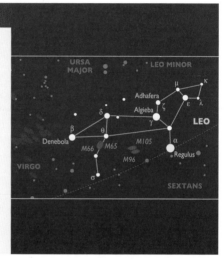

One of the most easily visualized constellation shapes, Leo the
Lion figured as one of the 13 original zodiacal constellations
established by the Babylonians.

KEY FACTS

The kingly constel-
lation Leo has 12
stars, with 8 of them
forming the asterism
known as the Sickle,
which represents the
lion's uplifted head.

+ months best
viewed: **March–April**

+ seasonal chart
location: **Spring,
center of chart**

+ alpha star:
Regulus

The ancient Egyptians revered the constellation Leo.
They associated it with the sun, and may have modeled
the Sphinx on its form. Leo also links to Hercules myths as
a representation of the Nemean lion that the hero choked
to death in the first of his 12 labors. The constellation
harbors spiral galaxies M65 and M66, which can be seen
with binoculars. Leo conveniently appears at the end of an
imaginary line formed from Polaris
and through the bowl edge of
the Big Dipper. Leo is also the
location of the Leonid meteor
showers, which are identified with
the comet Tempel–Tuttle and peak
in mid-November.

Leo Minor
The Small Lion   Size on the sky:

Leo Minor requires a dark night for visibility, but it lies almost directly overhead in the mid-northern latitudes in early spring.

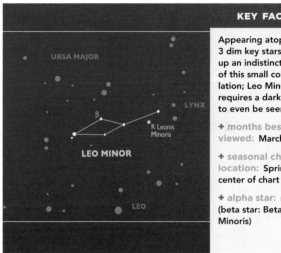

KEY FACTS

Appearing atop Leo, 3 dim key stars make up an indistinct shape of this small constellation; Leo Minor requires a dark night to even be seen.

+ months best viewed: **March–April**

+ seasonal chart location: **Spring, center of chart**

+ alpha star: **none (beta star: Beta Leonis Minoris)**

Like other 17th-century additions to the celestial charts, Leo Minor lacks a lot of supporting documentation or mythology. Polish astronomer Johannes Hevelius isolated the formation, which lacks a labeled alpha star, probably as the result of an oversight. Beta star Leonis Minoris is not the constellation's brightest; that distinction belongs to R Leonis Minoris, an oscillating star with more than a five-point range of magnitude during the course of a year. The ancient Egyptians recognized the small Leo Minor group of stars as the hoofprints of a herd of gazelles in flight from neighboring big cat Leo.

||||||||||||||||||||||||||||||||||||||||||||||||||||||||||||||||||||||||||||||||||

Lepus

The Hare   Size on the sky: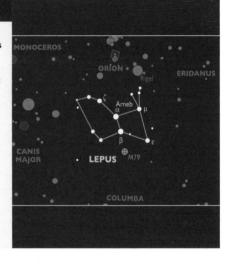

Just as Orion keeps close his celestial hunting dogs, he also has his prey nearby in the form of Lepus, the small constellation representing the Hare.

KEY FACTS

The constellation Lepus has 11 key stars that make the small hare shape, including the somewhat distinct ears, which appear just below Rigel in constellation Orion's left foot.

+ **months best viewed:** January–February

+ **seasonal chart location:** Winter, southeast quadrant

+ **alpha star:** Arneb

Lepus the Hare sits crouched and ready to flee just west of Sirius, the bright star in the constellation of a pursuer, the dog Canis Major. Midwinter marks Lepus's highest position in the sky, but it remains close to the horizon for viewers in mid-northern latitudes and dips down below it for part of the year. Lepus contains the globular cluster M79, unusual for the fact that it sits within a southern constellation. Most other globular clusters are found at the center of the Milky Way, which lies a significant distance away. It is possible that globular cluster M79 was formed in the neighboring Canis Major dwarf galaxy.

||||||||||||||||||||||||||||||||||||||||||||||||||||||||||||||||||||||||||||||||||||||

Libra

The Scales   Size on the sky:

Libra falls within the constellations of the zodiac, the only one named for an inanimate object. It appears between zodiac neighbors Scorpius and Virgo.

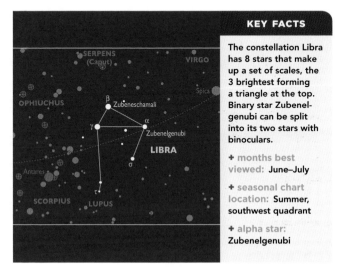

KEY FACTS

The constellation Libra has 8 stars that make up a set of scales, the 3 brightest forming a triangle at the top. Binary star Zubenel-genubi can be split into its two stars with binoculars.

+ months best viewed: June–July

+ seasonal chart location: Summer, southwest quadrant

+ alpha star: Zubenelgenubi

A constellation of the northern summer sky, Libra lacks star magnitude when compared to its neighbors. Arab astronomers, and the Greeks before them, considered the alpha and beta stars of Libra to be the claws of the scorpion of neighboring Scorpius. Their names reflect this designation: Zubenelgenubi means "southern claw" and Zubeneschamali means "northern claw." Later, constellation Libra's scales were associated with those held by Astraea, the Greek goddess of justice. Libra's brightest stars can be found along a line traced from alpha star Antares in Scorpius to alpha star Spica in Virgo.

Lynx

The Lynx   Size on the sky:

The constellation Lynx came into being in the late 17th century to fill an empty celestial area between Ursa Major and Auriga. Polish astronomer Johannes Hevelius introduced it.

KEY FACTS

The 7 key stars of this inconspicuous constellation form a kinked line somewhat suggestive of a rearing lynx, located to the north of stars Castor and Pollux in Gemini.

+ **months best viewed:** February–March

+ **seasonal chart location:** Winter, northeast quadrant

+ **alpha star:** Al Fahd

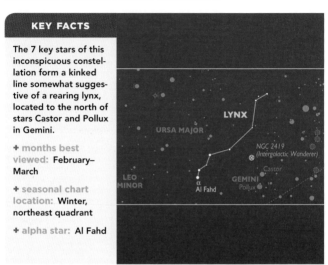

Inconspicuous constellation Lynx makes a difficult sighting for casual stargazers. It is said that Lynx earned its name because it requires the observer to possess the sharp eyesight of the northern forest cat to spot it. As a kind of compensation, the constellation contains a number of double stars to reward the observer who uses even a modest telescope. Lynx also harbors a deep sky object in the form of the Intergalactic Wanderer (NGC 2419). The name "Wanderer" refers to the belief that the object is so far from the core of the Milky Way that it might be expected to break away from its current loose orbit around the galaxy.

Lyra
The Lyre   Size on the sky: 👆

Alpha star Vega steals the show in constellation Lyra, appearing directly overhead in the mid-latitude summer sky. It is ranked the third brightest star in the sky.

KEY FACTS

The 6 stars of the easily identified constellation Lyra compose the ancient musical instrument; its alpha star Vega ranks among the brightest in the sky.

+ **months best viewed:** July–August

+ **seasonal chart location:** Summer, center of chart

+ **alpha star:** Vega

Lyra, the Lyre, belonged to Orpheus, Apollo's tragic son. He lost his beloved wife, Eurydice, not once but twice, when he blew the chance to retrieve her from Hades. Remaining faithful to her memory, Orpheus was killed by a group of scorned women. An impressed Zeus then sent Orpheus's lyre to the sky. With Deneb in Cygnus and Altair in Aquila, alpha star Vega in Lyra forms an asterism known as the Summer Triangle. In 12,000 or so years, alpha star Vega is slated to become the polestar. The Ring Nebula (M57), located at the edge of Lyra, is formed of gas, and it is visible through a telescope.

||||||||||||||||||||||||||||||||||||||||||||||||||||||||||||||||||||||||||||||||||||

Monoceros

The Unicorn   Size on the sky:

The name "Monoceros" derives from the Greek for "one horned." A constellation created in the 17th century, the Unicorn was first officially recorded by German astronomer Jakob Bartsch.

KEY FACTS

Its 8 stars of fairly low magnitude compose a dim, stylized unicorn with its "head" stars close to Betelgeuse. Brighter neighboring stars and features help locate Monoceros.

+ months best viewed: **January–February**

+ seasonal chart location: **Winter, southeast quadrant**

+ alpha star: **Ctesias**

Monoceros appears within the asterism known as the Winter Triangle, which is composed of Betelgeuse in Orion, Procyon in Canis Minor, and Sirius in Canis Major. The constellation is dim in comparison to those very bright stars, but it fills an important gap on the sky chart, and its position on the Milky Way provides many good reference points. One highlight—open star cluster M50—appears between Procyon and Sirius and can be spotted through binoculars. Monoceros may refer to a biblical creature that decided to play in the rain rather than join Noah and the other animals on the ark to escape the great flood.

|||||||||||||||||||||||||||||||||||||||||||||||||||||||||||||||||||||||||||||||||||||||

Ophiuchus

The Serpent Bearer *Size on the sky:*

Southern constellation Ophiuchus interrupts neighboring Serpens, splitting it in two. Entwined mythologies perpetually link them.

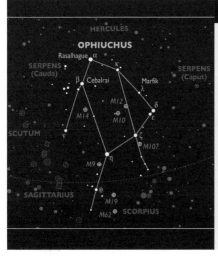

KEY FACTS

The constellation Ophiuchus has 14 stars that make up the blocky figure of a man. The Serpent Bearer's alpha star, Rasalhague, represents the head.

+ months best viewed: **June–July**

+ seasonal chart location: **Summer, center of chart**

+ alpha star: **Rasalhague**

The constellation Ophiuchus likely honors Asclepius, the Greek god of medicine. He appears between the head and tail of Serpens the snake, who taught the god about the healing properties of plants. Asclepius used this knowledge to raise a fatally wounded Orion. A concerned Hades, god of the dead, fearing irrelevance, had Zeus kill Asclepius. Now healer and serpent watch from the sky, easily identified just to the west of the Milky Way. The constellation's location makes it rich in deep sky objects, including a series of globular clusters—M9, M10, M12, M14, M19, M62, and M107—all visible with binoculars.

Orion

The Hunter   Size on the sky:

Perhaps the most easily recognized constellation in the sky, Orion takes center stage on the celestial equator with more than its fair share of superbright stars.

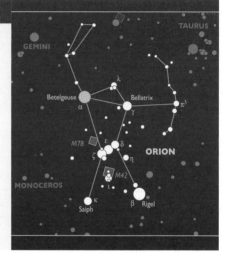

KEY FACTS

Some 20 stars make up the obvious figure of a hunter. The constellation is punctuated with huge stars of high magnitude such as Betelgeuse, Rigel, and Bellatrix.

✦ months best viewed: January–February

✦ seasonal chart location: Winter, southeast quadrant

✦ alpha star: Betelgeuse

Without a doubt, Orion is a stellar rock star. Among many superlatives, three very bright stars form the hunter's iconic belt; red-tinted, variable Betelgeuse boasts a diameter 300 to 400 times wider than the sun's; and blue-white supergiant Rigel shines 57,000 times brighter than the sun. The great hunter of Greek mythology, Orion appears in the vicinity of his canine companions Canis Major and Minor and other associates, such as Lepus the Hare. Located below Orion's belt, the Great Orion Nebula (M42) is visible as a cloud patch to the naked eye. Its gas cloud churns out new stars at a furious pace.

Pegasus

The Winged Horse Size on the sky:

Pegasus shares the sky with other constellations attached to the same mythology: Andromeda, Cassiopeia, Cepheus, Cetus, and the heroic Perseus.

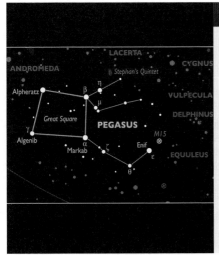

KEY FACTS

The horse figure in this large constellation is made up of 15 stars, with 4 of them forming the Great Square of Pegasus, a prominent asterism.

+ months best viewed: September–October

+ seasonal chart location: Autumn, center of chart

+ alpha star: Markab

Constellation Pegasus is rich in deep sky objects, including the M15 cluster and Stephan's Quintet, a group of five galaxies, four of which are interacting with one another. Pegasus also is a fertile source of exoplanets, planets that circle stars other than the sun. Pegasus shares a star, Alpheratz, and a story with constellation Andromeda (see page 116). Alpheratz forms part of the Great Square asterism with three stars from Pegasus. Pegasus also has associations with the hero Bellerophon, who fought the Chimera with the help of the steed. He had tamed the flying horse with aid from the goddess Athena.

Perseus
The Hero   Size on the sky:

Constellation Perseus, named for a superhero of Greek mythology, has a lot going for it: many bright and easily spotted stars, a superlative meteor shower, and a fascinating double star cluster.

KEY FACTS

Perseus has 14 key stars, with 6 of them at magnitude 3 or lower; the constellation straddles the Milky Way between associated formations Cassiopeia and Andromeda.

✦ **months best viewed:** November–December

✦ **seasonal chart location:** Autumn, northeast quadrant

✦ **alpha star:** Mirfak

Perseus belongs to the same mythological adventure as neighboring constellations Andromeda, Cassiopeia, Cepheus, Cetus, and Pegasus. The hero rescued captive Andromeda, daughter of Cassiopeia and Cepheus, from the predatory sea monster Cetus. The constellation is prominent in the late autumn sky and contains bright and notable signposts. Among them is Algol, the beta star, an eclipsing variable that dims by a full magnitude every three days. Perseus contains double cluster NGC 869 and NGC 884, two separate groups of stars that can be viewed through binoculars. It also is point of origin for the mid-August Perseid meteor showers.

|||||||||||||||||||||||||||||||||||||||||||||||||||||||||||||||||||||||||||||||||||||||

Pisces

The Fish   Size on the sky:

The fish of Pisces are on the small side, but the cord joining them by their tails spans a wide area and crosses the ecliptic, the apparent path taken by the sun across the sky.

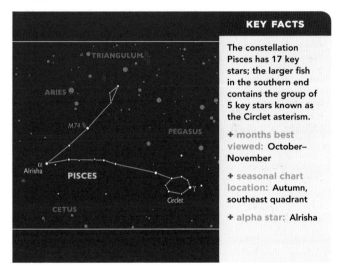

KEY FACTS

The constellation Pisces has 17 key stars; the larger fish in the southern end contains the group of 5 key stars known as the Circlet asterism.

+ **months best viewed:** October–November

+ **seasonal chart location:** Autumn, southeast quadrant

+ **alpha star:** Alrisha

Pisces is an ancient constellation of the zodiac that forms a large V in the sky. Its southern asterism, known as the Circlet, appears just south of the Great Square of Pegasus. The outside cord attached to the smaller fish is galaxy M74, a faint but elegant spiral galaxy whose arms can be resolved with an 8-inch-aperture telescope. For the ancient Greeks, the two fish of Pisces represented the goddess Aphrodite and her son Eros, who both changed themselves into fish to escape the sea monster Typhon, often equated with Cetus. The cord, tied at alpha star Alrisha, holds mother and son together as they swim.

Piscis Austrinus

The Southern Fish   Size on the sky:

The key to locating this low southern constellation—low to the horizon even at its highest for northern mid-latitude observers—is its formidable alpha star Fomalhaut.

KEY FACTS

Numerous ancient cultures have visualized these 10 stars as a fish; the name of the alpha star, Fomalhaut, is derived from the Arabic for "mouth of the fish."

+ **months best viewed:** September–October

+ **seasonal chart location:** Autumn, southwest quadrant

+ **alpha star:** Fomalhaut

One of the original 48 constellations Ptolemy identified, Piscis Austrinus is located in the "watery" section of the sky just south of Aquarius and east of Capricornus. In Greek mythology, the Southern Fish was considered to be the sire of the two fish in the constellation Pisces. Alpha star Fomalhaut, considered a young star at between 100 and 300 million years old, ranks as the 18th brightest star to the naked eye and is a mere 25 light-years from Earth. In 2008, the Hubble telescope photographed a planet orbiting Fomalhaut. Named Fomalhaut b, it was the first confirmed extrasolar planet detected via direct imaging.

Puppis

The Stern   Size on the sky:

Puppis is one of the three breakout constellations created in 1763 from Argo Navis, one of the original constellations on Ptolemy's list of 48.

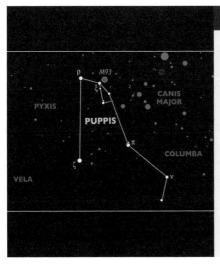

KEY FACTS

The constellation Puppis has 12 stars known as the stern of the largest part of the ship *Argo*; it lies in the rich starfields of the Milky Way.

+ months best viewed: February–March

+ seasonal chart location: Winter, southeast quadrant

+ alpha star: none (brightest: Zeta Puppis)

Puppis, disappointingly to some, doesn't represent a cute little dog, but the stern of the mythological ship *Argo*. The *Argo* carried Jason and the Argonauts on the quest to steal the Golden Fleece, once sported by Aries the Ram, from the clutches of a dragon. When Argo Navis was dismantled during the constellation update of the modern era, its stars were not relabeled within the new constellations and so Puppis was left without a labeled alpha star. Its brightest star is Zeta Puppis, also known as Naos, from the Greek for "ship." A blue supergiant, Zeta Puppis is a spectacular naked-eye star of the second magnitude.

||||||||||||||||||||||||||||||||||||||||||||||||||||||||||||||||||||||||||||||||||||||||||||||||

Sagitta
The Arrow   Size on the sky: 👍

Although the shapes associated with many constellations often stretch the imagination, the arrow-like form of the small but distinct Sagitta cleanly hits the mark.

KEY FACTS

The unmistakable arrow shape of the constellation Sagitta contains 5 stars. The arrow's shaft contains M71, a globular cluster visible through binoculars.

+ months best viewed: August–September

+ seasonal chart location: Summer, southeast quadrant

+ alpha star: Sham

The third smallest constellation, Sagitta lies totally within the Summer Triangle asterism formed of Altair in Aquila, Deneb in Cygnus, and Vega in Lyra. Most ancient cultures made associations to the arrow, and there are a number of arrow stories to choose from to explain Sagitta's celestial presence. In Greco-Roman myth, it could be the arrow Cupid (Eros) used to spread the love, or the arrow Apollo used to slay the Cyclops, or the one Hercules deployed to dispatch the Stymphalian birds in one of his 12 labors. A relationship to Sagittarius, despite the name "arrow," is not widely accepted.

||||||||||||||||||||||||||||||||||||||||||||||||||||||||||||||||||||||||||||

Sagittarius

The Archer   Size on the sky: 🖐

Spread across a wide band of the Milky Way, Sagittarius is a treasure trove of celestial superlatives, including two prominent asterisms and eight high-magnitude stars.

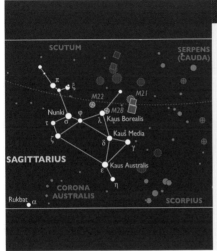

KEY FACTS

The complex figure of the constellation Sagittarius contains 22 stars, as well as the clearly defined Teapot asterism and the Milk Dipper asterism.

+ **months best viewed:** July–August

+ **seasonal chart location:** Summer, southeast quadrant

+ **alpha star:** Rukbat

Appearing south of Vega, the Teapot asterism contains the eight central stars of Sagittarius. The Milk Dipper includes the ladle and lambda star handle that form the Teapot's handle. Sagittarius's alpha star is not its brightest; that would be epsilon Kaus Australis. Sigma star Nunki, the second brightest, appeared on the Tablet of the 30 Stars of the ancient Babylonians. The centaur of Sagittarius—half man, half horse—pursues his supposed prey, the neighboring constellation Scorpius, through the sky. Sagittarius is often identified with Chiron of Greek mythology, who paradoxically was a wise and peaceful creature.

III

Scorpius

The Scorpion   Size on the sky: 🖐️

The narrow outline of Scorpius harbors 2 of the 25 brightest stars as well as outstanding star clusters, including the spectacular Butterfly cluster (M6).

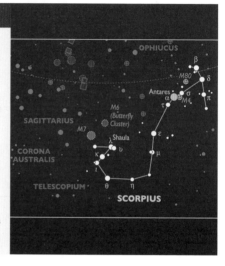

KEY FACTS

A recognizable scorpion shape is made from 17 main stars in the constellation Scorpius, anchored by bright Antares in the core and Shaula at the tail.

+ **months best viewed:** July–August

+ **seasonal chart location:** Summer, southeast and southwest quadrants

+ **alpha star:** Antares

A straightforward constellation, Scorpius is located by first-magnitude star Antares, known to the Romans as Cor Scorpionus, or "heart of the scorpion." A red supergiant, about 300 times as large as the sun, Antares's name means "rival of Mars," alluding to its size and color. The Butterfly cluster (M6), one of many open and globular star clusters in the vicinity, reveals its butterfly shape through binoculars. In Greek mythology, Scorpius is tied to the constellation Orion as the animal that inflicted the fatal wound on the hero's leg. The two constellations appear on opposite sides of the sky to keep them separated.

Scutum

The Shield Size on the sky: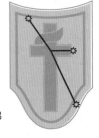

This faint modern constellation was created by Polish astronomer Johannes Hevelius in the late 17th century in the feature-rich Milky Way.

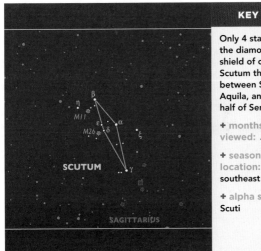

KEY FACTS

Only 4 stars compose the diamond-like shield of constellation Scutum that appears between Sagittarius, Aquila, and the lower half of Serpens.

+ months best viewed: **July–August**

+ seasonal chart location: **Summer, southeast quadrant**

+ alpha star: **Alpha Scuti**

Scutum lacks any bright or otherwise notable stars, but its location on the Milky Way puts it in close proximity to several open star clusters, including M11, the so-called Wild Duck cluster. Binoculars alone will reveal this formation that resembles a dense flock of waterfowl, and an 8-inch telescope will show thousands of glittering stars. The Wild Duck appears just southeast of Scutum's northernmost star. The constellation's original name was Scutum Sobiescianum, or "shield of Sobieski," honoring Polish King (and Hevelius's patron) John III Sobieski, who defeated the Ottoman Empire in the 1683 Battle of Vienna.

Serpens

The Serpent   Size on the sky: 🤚🤚🤚

The only split constellation in the night sky, Serpens is divided between head and tail, winding over the shoulder of Ophiuchus, the Serpent Bearer, in the middle.

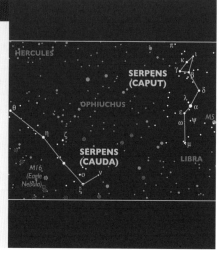

KEY FACTS

Two groups of stars, the triangular head (Caput) and elongated tail (Cauda), make up the two separate parts of the serpent constellation Serpens.

+ months best viewed: **June–July**

+ seasonal chart location: **Summer, southern half of chart**

+ alpha star: **Unukalhai**

The constellation Serpens formerly encompassed the whole scenario of snake head, tail, and Ophiuchus. They were physically and mythologically united, representing the Greek god of medicine, Asclepius, and the serpent who taught him about the healing power of plants. Serpens Cauda contains the combination nebula and star cluster called the Eagle Nebula (M16), which occurs in an active region of star formation about 7,000 light-years away. Both the nebula and cluster can be spotted with an 8-inch telescope. M16 was famously photographed by the Hubble telescope, which captured a prominent area of star destruction and formation.

|||||||||||||||||||||||||||||||||||||||||||||||||||||||||||||||||||||||||||||||||||||||

Taurus
The Bull   Size on the sky: ✋

Taurus, a zodiacal constellation, is easily located because nearby Orion's belt points directly northwest to Aldebaran, the Bull's alpha star.

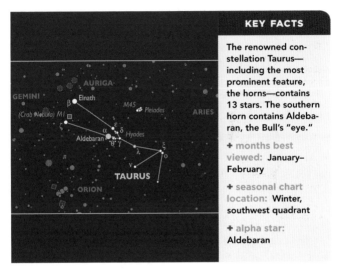

KEY FACTS

The renowned constellation Taurus—including the most prominent feature, the horns—contains 13 stars. The southern horn contains Aldebaran, the Bull's "eye."

+ months best viewed: **January–February**

+ seasonal chart location: **Winter, southwest quadrant**

+ alpha star: **Aldebaran**

The Egyptians associated Taurus with Osiris, god of life and fertility. Greek god Zeus disguised himself as a bull to capture Europa and carry her to the continent that bears her name. Taurus contains two prominent star clusters, the Pleiades and the Hyades. The Pleiades (M45), an open cluster, is the more famous and displays six or seven of its hundreds of stars to the naked eye. Vying for celebrity is M1, the Crab Nebula. It was the first deep sky object cataloged by Charles Messier, in 1758. Cultures from the American Southwest to China witnessed and recorded the death of the star in 1054 that created the nebula.

||||||||||||||||||||||||||||||||||||||||||||||||||||||||||||||||||||||||||

Triangulum

The Triangle   Size on the sky:

Despite its nonmythological, utilitarian name—a feature of many of the modern constellations—Triangulum was on Ptolemy's list of the original 48.

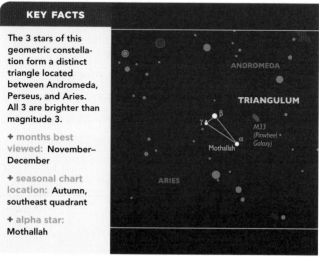

KEY FACTS

The 3 stars of this geometric constellation form a distinct triangle located between Andromeda, Perseus, and Aries. All 3 are brighter than magnitude 3.

+ months best viewed: November–December

+ seasonal chart location: Autumn, southeast quadrant

+ alpha star: Mothallah

In late autumn, Triangulum is easy to find just east of Andromeda. The Pinwheel galaxy (M33) lies west of the constellation, a spiral galaxy like our own Milky Way. Under good viewing conditions, it appears like a slight glow, although a substantial telescope is needed to view the pinwheel effect. The ancient Hebrews appear to have named this constellation for the small percussion instrument, and the name stuck. The Greeks also took notice that these stars resembled their letter *delta*, and it was known for a time as Deltoton. The theme continues in the name for the alpha star, Mothallah, which means "triangle" in Arabic.

Ursa Major

The Great Bear Size on the sky:

To various cultures, the formidable assemblage of stars in Ursa Major represented a bear, a chariot, a horse and wagon, a team of oxen, and even a hippopotamus.

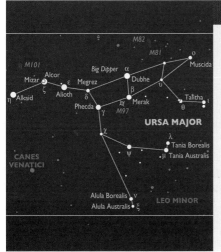

KEY FACTS

The complex figure of the constellation Ursa Major contains 20 stars, including the 7-star asterism of the Big Dipper, representing the bear's rear torso and tail.

+ **months best viewed:** March–April

+ **seasonal chart location:** Spring, center of chart

+ **alpha star:** Dubhe

During the year, prominent Ursa Major seems to run in a circle with its back to Polaris, the North Star. The constellation's many named stars include Alcor and Mizar, which appear to be a binary system but are not. Galaxies M81 and M82 lie close together and may be in collision with each other. In Greek mythology, the Great Bear represents Callisto, yet another victim of the goddess Hera's jealousy on finding that her husband, Zeus, had strayed again. When Callisto's son mistakenly tried to kill her while hunting, Zeus placed them both in the sky to protect them.

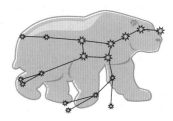

Ursa Minor
The Little Bear   Size on the sky: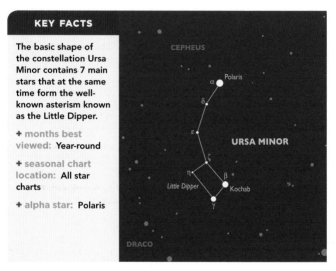

The Little Bear plays second fiddle to none, containing the North Star, Polaris, as its alpha star. The constellation and its polestar are visible year-round.

KEY FACTS

The basic shape of the constellation Ursa Minor contains 7 main stars that at the same time form the well-known asterism known as the **Little Dipper.**

✦ months best viewed: **Year-round**

✦ seasonal chart location: **All star charts**

✦ alpha star: **Polaris**

Navigators, wanderers, and skygazers throughout history have found their way by the enduring presence of Ursa Minor and Polaris, indicating the celestial pole. This won't always be the case. Over millennia, the celestial pole shifts. It will be closest to Polaris around 2100, and then will start to move away, passing other stars and eventually settling at Vega for a while in about 12,000 years. Ursa Minor appears in the sky as the mythological child of the Great Bear, who represents Zeus's paramour Callisto. Late December brings the Ursid meteor shower, which displays about ten meteors an hour at its peak.

Virgo
The Virgin   Size on the sky:

The only zodiacal constellation representing a woman, Virgo is the second largest constellation and a fertile area for viewing star clusters and galaxies.

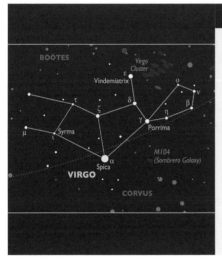

KEY FACTS

The figure of a seated woman in the constellation Virgo contains 13 main stars, with bright alpha star Spica in the vicinity of her left hand.

+ **months best viewed:** May–June

+ **seasonal chart location:** Spring, southeast quadrant

+ **alpha star:** Spica

Mythologically, Virgo appears to represent a goddess. For many she was Dike, the goddess of justice, or possibly Demeter, the goddess of agriculture. Virgo's bright alpha star, Spica, is a useful point of orientation as part of an alliterative mnemonic, "arc to Arcturus, then speed on to Spica." This directs one's gaze south from the end of the Big Dipper's handle to find Arcturus in Boötes, then south again to Spica just below it. Northwest of Virgo is an area dense in star clusters and galaxies—thousands of them. To the southwest is the Sombrero galaxy (M104), with a dark band across its middle that serves as the brim.

Further Resources

BOOKS

David, Leonard. *Mars: Our Future on the Red Planet*. National
Geographic Partners, 2016.

Kaufman, Marc. *Mars Up Close: Inside the Curiosity Mission*.
National Geographic Society, 2014.

Ridpath, Ian, and Wil Tirion. *Stars and Planets,* 4th ed. Princeton
Field Guides. Princeton University Press, 2007.

Trefil, James. *Space Atlas: Mapping the Universe and Beyond*.
National Geographic Society, 2012.

NATIONAL & INTERNATIONAL ORGANIZATIONS

European Space Agency
www.esa.int

Harvard-Smithsonian Center for Astrophysics
www.cfa.harvard.edu

Hubble Space Telescope
www.hubblesite.org

International Astronomical Union
www.iau.org

National Aeronautics and Space Administration
www.nasa.gov

Space Weather Prediction Center
www.swpc.noaa.gov

PERIODICALS

Astronomy magazine
www.astronomy.com

Sky & Telescope magazine
www.skyandtelescope.com

WEBSITES

Ask an Astronomer, http://curious.astro.cornell.edu
Site run by Cornell University where users can post questions
for Cornell astronomers.

EarthSky, www.earthsky.org
A science website that provides daily updates on Earth,

space, and phenomena in the human world.

Kids Astronomy, www.kidsastronomy.com
> Free astronomy website designed to teach children about the universe, run by KidsKnowIt Network.

Moon Connection, www.moonconnection.com
> Detailed website about the moon, its phases, and missions to the moon.

Nine Planets, www.nineplanets.org
> A detailed overview of the planets, moons, and other objects in our solar system.

Planetary Society, planetary.org
> A community that empowers citizens to advance space science and exploration, founded by Carl Sagan and helmed by CEO Bill Nye.

Space.com, www.space.com/news
> Multimedia website dedicated to space news.

Star Date, stardate.org
> The public education arm of the University of Texas McDonald Observatory.

About the Author

CATHERINE HERBERT HOWELL, a former National Geographic staff member, has written extensively on natural history. She explored the relationships between people and plants in *Flora Mirabilis: How Plants Have Shaped World Knowledge, Health, Wealth, and Beauty* (2009) and covered the importance of birds in culture in the *National Geographic Bird-watcher's Bible* (2012). Howell serves as a master naturalist volunteer in Arlington, Virginia.

Credits for sky and constellation charts and moon map
Moon mosaic and typography: pp. 38-39, Lunar Reconnaissance Orbiter (LRO), NASA, Arizona State University; seasonal night sky charts: pp. 108-115, Wil Tirion; constellation charts and mythological icons: pp. 116-171, Brown Reference Group.

ILLUSTRATIONS CREDITS

Index

Boldface indicates
illustrations.

Since 1888, the National Geographic Society has funded more than 12,000 research, exploration, and preservation projects around the world. National Geographic Partners distributes a portion of the funds it receives from your purchase to National Geographic Society to support programs including the conservation of animals and their habitats.

National Geographic Partners
1145 17th Street NW
Washington, DC 20036-6688 USA

Become a member of National Geographic and activate your benefits today at natgeo.com/jointoday.

For information about special discounts for bulk purchases, please contact National Geographic Books Special Sales: specialsales@natgeo.com

For rights or permissions inquiries, please contact National Geographic Books Subsidiary Rights: bookrights@natgeo.com

ISBN: 978-1-4262-1785-2

Printed in Hong Kong

16/THK/1

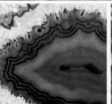

PACK WHAT MATTERS
and explore with the experts!
Slim ● Affordable ● Easy-to-use field guides

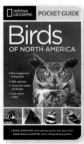

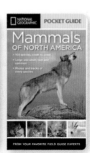
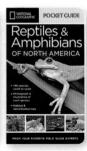

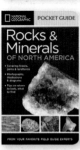
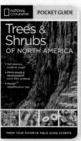